T0139033

Myth, Chaos, and Certainty

Myth, Chaos, and Certainty

Notes on Cosmos, Life, and Knowledge

Rosolino Buccheri

JENNY STANFORD
PUBLISHING

Published by

Jenny Stanford Publishing Pte. Ltd.
Level 34, Centennial Tower
3 Temasek Avenue
Singapore 039190

Email: editorial@jennystanford.com
Web: www.jennystanford.com

British Library Cataloguing-in-Publication Data
A catalogue record for this book is available from the British Library.

Myth, Chaos, and Certainty: Notes on Cosmos, Life, and Knowledge

ISBN 978-981-4877-33-6 (Hardcover)
ISBN 978-1-003-08869-1 (eBook)

To my beloved grandsons,

Rosolino and Aldo

Contents

Foreword

Whoever will read this Rosolino Buccheri's essay will believe himself to be sailing to a known sea, since he would glimpse along a coast apparently known villages and reference points that he will later recognize to be different from what he had in his mind. Scrolling through the pages, in fact, we will find elements of which we knew perhaps the existence but never understood their interdependence. Across the text we will find moments of the rational tradition, mythic elements, physical and biologic concepts, and traits of the sociologic and anthropologic knowledge. But, above all, we will find unusual connections.

It is certainly true that many authors, along the development of the Western thought, have tried to propose a comparison between what science tells us about the evolution of the universe and of life and about how the development of the social behavior appears to us. However, the typicality with which Buccheri deals with such a very risky topic is new and fresh, due also to his peculiar formation.

Many of them who have faced it have ended up producing a reading that wanted to be truthful. Naively, they did not understand that any interpretative proposal of the social behavior cannot be other than conjectural and provisional. Buccheri does not run such a risk. Manifestly or latently but gamely, he continuously warns, stimulates, suggests hypotheses or provisional interpretations, as it must be done when facing such risky themes.

Evolution of the universe, biologic evolution, and knowledge evolution. Evidently, the kind of evolution is not the same in the three cases, so it is for the related "laws." The laws that underlie the path bringing to the known cosmos from the *Big Bang*, are they the same as those describing the path from the *Last Universal Common Ancestor* (LUCA) to the biodiversity, instantiated in the many thousands of species today coexisting? Are they the same as those labeling the way going from the stories full of onomatopoeic sounds of the first hominids to the construction of the most ancient myths and to the realization of the powerful formal representations of natural processes, both physical or biologic?

The literature, furthermore, makes a distinction between two views about nature: the *formaliter spectate* nature, that is, the nature described by intellectual categories (not necessarily only those, indicated by Immanuel Kant), and the *materialiter spectata* nature, that is, consisting in the totality of the phenomena; the second is the one convincingly addressed by the author.

The need for disambiguation involves also the complexity. The complexity discussed in biology—for example, the complexity of pathologies like cancer or the complexity of molecular networks characterizing all living beings—is the same complexity of a global market or that involved in the social interactions within a South Tyrolean or Berber community.

Being well equipped with scientific and humanistic knowledge, Buccheri shows a lot of caution in all the discussions, extricating himself from the tangle of the faced problems. For this, reading this book will be fascinating for those who think that intellectual eclecticism and the thought that progresses for a similarity of ideas are praiseworthy features. Reading this work is a polymorphic adventure worthy to be lived with an open mind and the availability to listen.

Giovanni Boniolo
Dipartimento di Scienze Biomediche e Chirurgico Specialistiche
Università di Ferrara
Ferrara, Italy

Introduction

This book tells us about the results of a long-time-pondered research, partially derived by the re-elaboration of some published essays,[1] partially by notes still hidden within dozen of folders. Essays and notes originally conceived with the aim of discussing the social behavior, looking in particular for the why of the observed irreconcilable differences of opinions and views, especially as it happens today where a confused tangle of radical movements gives rise to conflicts and clashes of any kind, capable to destroy long-time-established communities' cohesion. Conflicts provoking fragmentation that overflows any wish of unity, for which every small community—often every single person—strongly feels a powerful need to distinguish himself by amplifying any small difference and by obscuring any important similarity. This reveals the rise of a haughty self-referencing attitude, generally supported by the acritical acquisition of any information, no matter if it is true but if it is good enough to please our own a priori beliefs. This phenomenon is excessively amplified today by the availability of the web's shop-window that, by widely distributing stereotypes, prejudices, and unverifiable information stated as true, ends with obstructing the necessary exchange of ideas, traditions, and beliefs between interacting peoples, thus supporting a dangerous radicalism portending disasters, as chronicles of the world tell us every day.

Such a situation, when watched from a wider perspective, recalls the necessity to look at the much more general problem of the evolution of human societies along the development of the knowledge.

The analysis of such a problem, being myself a common citizen who lives and looks at his time from his particular observation point, is inevitably affected by the deforming lens of my professional

---

[1]Some of these works have been produced along a fruitful editorial and research activity in the framework of AKOUSMATA·*orizzonti dell'ascolto*, others published by the *Centro Internazionale di Studi sul Mito*, and still others derived by personal notes and articles, some of which published between 2016 and 2017 on the online magazine *Dialoghi mediterranei*.

working ambit—astrophysics. Through this deforming lens, I got the suggestion to compare the behavior of the nature, as described by the sciences, with man's social behavior, starting from the ascertainment that both in the human behavior and in all natural occurrences, particularly in the magnificent phenomena occurring in the immensity of the cosmos, we always observe a swinging equilibrium between the opposite tendencies to fragmentation and cohesion. These considerations led me to concretely elaborate under a unifying light all those, apparently detached, aspects that characterize the variety of human behavior as well as the variety of the natural configuration. However, I took seriously into account, as an element of interpretative caution, the difference between the automatisms of the nature laws that do not allow any exception in their applications and the human responsibility inherent to the possibility to choose. A fundamental difference, this, for which the solution of any human behavior problems rests largely on our will, affected as it is by the need to decide whether to rely on our turmoil of feelings, often biased by interests or preconceptions,[2] or to open our mind to an interaction supported by the search for reliable sources of knowledge, aiming at a rationally coherent view and messenger of general utility and harmony. A will hopefully supported by the ability to highlight the need that man, along his way to knowledge, would not fall into the allurement to keep only the emotionally agreeable information, but that, as nature does with its rigorous laws, entrusts to rationality the task to run out every unverifiable/unverified data, even at the risk to select unpleasant ones. An attitude, therefore, apt to privilege the critical reasoning and willingness to listen to others—as it is done normally and with success within the scientific community—so as to widen our knowledge horizons, reduce fragmentation, and concur to the search for a highly needed consonance, which in everyday life is essential for a pacific and constructive coexistence.

By comparing the results of the natural laws acting in the evolution of the universe and of life with the human behavior along the process of knowledge, we discover two important elements that weaken the veil under which nature, following Heraclitus, hides itself, thus allowing us to extend toward ambits of a larger perspective the limits of the present discussion.

----

[2]No matter if ideologically rooted or determined by vital needs.

The first element consists in the fact that the description of both nature laws[3] and man behavior can be effactually schematized by means of two basic properties, opposite but in reciprocal collaboration. In the cosmos, the presence of well-defined laws directs their evolving run along directions from them determined but unknown to us, together with a continuous flowering of an enormous and apparently contrasting variety of configurations, which, too, contribute to the comprehension of the laws from which they derive. In the growth of life, between the formation of the organic molecules emerged from the complexification of the elementary matter and the appearance of humans, the variety of living beings is controlled by the need to fit to the Earth milieu. In human beings, the presence of a great variety of points of view and opinions, often in reciprocal contradiction but within an ordering principle dictated by reason, on average, addresses its path. Reason that reveals to be quite essential for orienting ourselves against that disarming self-referencing attitude whose fishing dock is just such a variety of opinions and points of view, an origin of misunderstandings, sometimes dramatically negative because of being associated to indestructible prejudices, intrinsic of all cultures. A reliance on reason is always necessary in order to effectively contribute to the social cohesion, and going to the best-possible comprehension of our physical and social setting.

The second element consists in the evident circularity of the three evolutions—universe, life, and knowledge—which tells us about the run-up, indefinite, and probably never conclusive, of the human being to the full knowledge of the nature from which he derives and of which he is continuously the producer.[4]

If we look at the evolution of each of the natural processes concerning matter—both inert and living—and knowledge, we

---

[3]The allocution "nature's laws" includes here, as previously mentioned, the laws describing the evolution of both the inert and the living matter. The reason will clearly appear later.

[4]It is worth noting that in the *Philosophy of Karl Popper*, Campbell supports the idea that "[. . .] the evolution—also in its biologic aspects—is a process of knowledge and [that] the paradigm of the natural selection explaining such increments of knowledge may be generalized to other epistemic activities, like learning, thought and science"— so, probably, just one evolution process including knowledge, the cosmos, and life (Campbell, 1974).

realize to be always in front of a "chaotic"[5] process, one of those processes characterized by an unforeseeable evolutionary direction, *apparently* random, even if in the simplest cases it can be described by exact mathematical equations. Characteristics, the last, that we usually find in *complex systems*, that is, in physical systems made up of a great number of mutually interacting elements, as in the cases here discussed.

Chaotic, in this sense, is the universe's evolution, as synthetically described in the first chapter of Part I. There, gravity, electromagnetic, and nuclear forces, together with other laws probably present at the start of the universe, are still pushing it toward a direction today unknown, but marked with an immense beauty, as shown by the myriad of forms and colors in which matter continuously transforms itself.

Chaotic is also the evolution of life—discussed in the second chapter of Part I—emerged from the evolution of inert matter, where chance addressed by necessity (to use Jacques Monod's words) gave rise to the enormous variety of living beings in our planet. Living beings that were originated by the spontaneous self-organization of inert matter, starting from its original elements born out of the nuclear fusion within stars. Matter ejected out—as will be explained in detail in the first chapter of Part I—and later subject to an evolutionary process characterized by a growing complexity and governed by specific mutations, which was able to produce organisms always more resistant to their milieu.

Chaotic is, finally, the evolution of man's knowledge—discussed in the third chapter of Part I. Here the cognitive duality, consisting in the coexistence of the unconscious and the rational working modalities of our brain, opposite but complementary, produces an uncontainable fragmentation of views and opinions, on average managed by rationality, when searching for a direction to move to. A direction tending toward an objectivity that can never be raised to

---

[5]The term "chaotic" is here meant in the mathematical sense and refers to the characteristics of physical systems with many parameters nonlinearly linked together, and therefore extremely sensible to even minimal variations of the initial conditions. A situation, this, that prevents the predictability of their subsequent states, as long as we look for their further run. Here, the concept of "chaos" is applied, for analogy, to the processes of life and of knowledge that, for their much greater complexity, are not certainly describable with precise mathematical equations. A circumstance, this, that confirms and increases the difficulty to foresee the evolution's direction.

absolute truth, but that can lead to a sharing of statements verifiable on Earth by whoever is willing to adhere to a widely accepted analysis methodology."

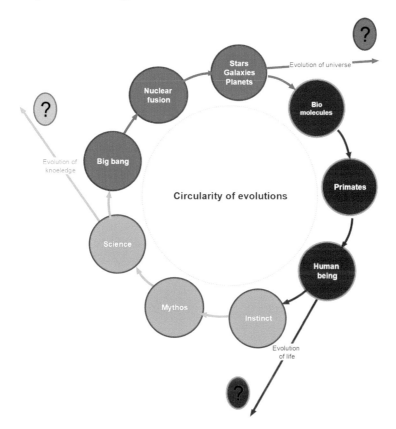

Circularity of evolution.

The clear, closed circularity connecting the three cited evolutions[6] demonstrates the theoretical-practical impossibility to formulate any final certainty on the content of our knowledge, affirming instead the idea of a virtuous circle of evolutions—guided by the combination, virtuous, too, of law and chance—where "every tale carries the print of other tales as much intricate and as much coherent."[7] Stories of stars, stellar systems, and other astonishing

---

[6]See the figure on this page.
[7]Bocchi and Ceruti (2006, p. 8).

phenomena, born, transformed, collapsed, and reborn under new forms, in the middle of dramatic and wonderful collisions and reciprocal destructions, triggered from the interlacing of casual fluctuations and physical laws by which they are severely controlled. Stories where new generations of stars enriched with material coming from the exhausted generations slowly modify the overall scenario toward a yet unknown future. Stories where life is originated in an ebullition of events, with the natural selection watching over casual mutations, able to adapt life to the environment and to lead it toward always more complex forms until the appearance of humans. Stories where, in the later evolution of the brain structure, the reason, able to understand the importance but contradictory *modus operandi* of the myth, may oversee it, embanking its negative effects in order to prevent any social marginalization so as to guide man toward a shared knowledge.

Circular stories without any start or end—stories where the springing of life and its evolution tracks the emergence and development of the universe until the appearance of human intelligence and the growth of man's cognitive and technical abilities, which, in turn, allow him to understand the evolution of the universe, his own origins, and his own story within it. Paraphrasing Chuang-Tzu, we could ask ourselves whether it is truly the universe from which a human being comes out or it is just the human being that generates the universe.[8]

Circular stories supported by a virtuous interaction between law and chaos, between *mythos* and reason—a sort of constant of nature that can be taken as a model in any social analysis in order to be able to deny any "Pindaric" flight by stiff utopias dreaming about ideal worlds characterized by a presumed "absolute" rationality. Utopias, far from having an absolute value since unavoidably watched by a multitude of diverse eyes and therefore contaminated by any yearnings or needs, physical, intellectual, or economical, thus implying fierce tearing disputes as clearly highlighted along the course of time.

Neither Hegel nor Marx then, neither only the "idea" nor only the observation of nature, but a continuous run-up between idea and

---

[8]Chuang-Tzu, a Chinese philosopher and mystic, founder of Taoism, together with Lao Tze, dreamed to be a butterfly and when he woke up, confused, was uncertain to be Chuang Tzu having dreamed to be a butterfly or to be a butterfly actually dreaming to be Chuang Tzu.

observation at the root of the origin and progressive evolution of the knowledge.[9] Evolution, therefore, that rises up to a basic principle for any aspect of our existence, a principle excluding any illusory presence of stability in the course of time, not only concerning cosmological events, but also regarding vital processes and, with reference to our brief path in the world, concerning also all those processes connected with the progress of the organization of human societies along the progress of knowledge.

Because of their large interdisciplinary implications, and due to the complexity of the topics involved, all considerations here proposed do not pretend to be exhaustive nor are they discussed with the necessary depth in all their facets, thus implying the obvious risk to fall into possible evaluation mistakes. There is no solution to such a deficiency, which could only be filled by the impossible presence of a lot of deep specialists, everyone an expert on one of the here touched disciplines, without, anyway, the grant to succeed to coherently connect such a large and diverse know-how.

As a partial justification about any possible failures, I recall with due humility a lovely booklet titled *What Is Life? The Physical Aspect of the Living Cell*, written by the Nobel Prize winner Erwin Schrödinger, one of the founders of quantum mechanics. Schrödinger, by discussing with his proverbial acuteness the connections between biology and physics concerning some problems of genetics, apologized for having introduced a topic of which he was not a specialist. At the same time, he wrote:

> "... to not see any way out other than try—although with a second hand knowledge and taking into account the risk to be mocked—a synthesis of facts and theories able to exit from the dilemma to have perceived the existence of a good quantity of elements useful to weld together all our knowledge in the field and, on the other side, to have understood the impossibility for only one mind to dominate more than a small specialized ambit."[10]

---

[9]In his introduction to *The Materialistic Conception of History*, by Marx and Engels, Fausto Codino writes, with reference to the scientific socialism theorized by Marx, "While materialism conceives nature as the only reality, in the Hegelian system it represents only the manifestation of the absolute idea and is therefore a sort of degradation of the idea. At any rate, in such a system the thought and its intellectual product, the idea, is the original element, nature exists only as derived by the idea" (Marx and Engels, 1966, p. 15).

[10]Schrödinger (1944).

In the past 70 or more years, the amount of detailed knowledge has grown out of proportion, thus making Schrödinger's dilemma much harder to solve than before, with the result of always new and greater difficulties to correctly evaluate any situation, without having the precise cognizance of the general context in which to place them. What we generally observe regarding this attitude manifests itself within two extremes. From one extreme, we see the inclination to almost ignore the context so as to engage ourselves in a deep work within our own sector of study, thus improving the ability to deal with any small details and, at the same time, to accept to remain closed within it. At the other extreme, we observe the attitude to argue about everything without having well evaluated any unavoidable connections, certainly in an apparently scholar way, but very often by ignoring an important but hidden part of the detail necessary to ensure the internal cohesion of the "global" analysis. It is surely very hard to successfully identify the correct details needed for a coherent "middle way" between the two extremes, and one may easily make errors. I am convinced anyway, as it was Schrödinger, that one should not be discouraged from trying a reasonable equilibrium between the two propensities.

With these premises, let me try my adventure, conscious as I am of its risks, which are for me much higher as they were for the great Schrödinger, both for the enormous distance separating me from such a giant of science and for the much larger commitment requested because of the past 70 years of knowledge increase since then. It is encouraging for me the certainty that persons more expert in each of the disciplines here will be able, by resuming these subjects, to adjust and correct them with much more care and deepness than I am doing here, both in general and in the details.

I have to stress, by the way, that everything written here is the result of only my readings and my personal experiences, and as a consequence, the result cannot be free from errors and prejudices. Prejudices that I tried to limit as much as possible by trying to imitate Johannes Kepler, when he in his *Somnium*, to confirm the veracity of Copernicus's heliocentric hypothesis, imagined observing Earth from the lunar surface. A mental condition, this, that allowed him to enlarge his field of view, avoiding at a large extent any emotional involvement, always portending prejudices, but also being aware that if from far we are able to see much ampler areas together with clearer reciprocal connections and cooler sensations, it is also true that it is possible to lose important details, whose voids could be filled with unconscious prejudices.

To dismiss at all any presumption of objectivity, let me add, finally, that all reasoning, however rigorous it may be on the logical side, it can never avoid containing unprovable assumptions, either external to the logical system used but necessary for the completeness of the reasoning[11] or for a possible lack of "logic solidity" of the system that prevents to foresee contradictory developments or undecided propositions.[12] It follows that a reasoning able to bring one to absolutely true conclusions does not exist—the same scientific method is able to bring us to a kind of truth that is shareable only within the framework of our daily experiences, provided that the specific facts that we are talking of may be experimentally verified by anybody on Earth.

All these considerations have created for me many perplexities when I started writing this book. The major one was produced from the circularity between man as a "product" and simultaneously "producer" of nature, a circumstance that forces me to turn around a circumference with an arbitrary starting point. In fact, the actual knowledge of the universe follows from the development of our culture and of the methods we use to unveil nature's secrets—knowledge about the universe that, in turn, emerged from the life evolution on Earth and culminated in the presence of man with the progress of his theoretical and technical skills, able to bring about discoveries, about himself and about the world in which he lives.

In the middle of such a dilemma, I decided to start with the evolution of the universe—a decision that could let people think to a "realistic" position from my part, considering that everybody thinks that the universe pre-exists everything else, a conception, by the

---

[11]To be remembered that Blaise Pascal was far invoking a collaboration between the *esprit de finesse* and the *esprit de geometrie*.

[12]Here I am implicitly recalling the famous "incompleteness theorems" on formal systems published by Kurt Gödel in 1931. Following these theorems, all formal systems, however complete they may be, can, in principle, produce propositions undecidable or not able to show their internal coherence (Nagel and Newman, 1968).

way, that may be connected to my profession of an astrophysicist. Actually, instead, the *raison d'etre* of such a decision, possibly implying an unavoidable *exophysical* position with respect to an *endophysical* one,[13] is not due to my intrinsic realism but only due to the pure necessity to choose *any* point of the circumference able to make more comprehensible the description, in coherence with the already cited Kepler's attitude to look at things from the high in order to minimize the emotional involvement with respect to particular intellectual preferences.

Therefore, after a brief but necessary synthesis on the evolution of the universe and of life until the appearance of humans, I could focalize my attention on the dynamics of the social interaction so as to discuss some hypotheses on the possible future of mankind, suspended between cohesion and fragmentation.

Beyond all possible evaluation errors, we have to be conscious of the temporariness of the present discussion. It depicts, in fact, a contingent situation of which we know only the past history but cannot have any knowledge of possibly unexpected future developments. If we take into account the chaotic nature of any evolutionary processes, we cannot ensure that the physical and the intellectual state of human beings, as we know it today, will ever be the same, characterized by identical properties. Personally, I have the tendency to exclude it. I don't think that man, as we know him nowadays, is already the apex of life evolution and that life could not further evolve toward other living structures, perhaps with greater knowledge abilities, if nothing else, for the irreducibility of history to a single precise path, independently of any new circumstance. Concerning knowledge, moreover, two possible, well different, directions could be envisaged. From one side a further increase in the existing fragmentation, leading us to a self-destroying Babel Tower is not impossible. In addition, I don't believe impossible

---

[13]The *exophysical* attitude, that is, the claim that man may be able to explore the world by looking from outside, and therefore assuming to be able to ignore at all his interactions with the outside, will be discussed in the premise of Part III. The terms "exophysics" and "endophysics" were coined by David Finkelstein in 1983 in order to distinguish physics seen from outside the world from physics seen from inside the world (Rössler, 1998, p. 27). Some insights have been discussed by Buccheri and Buccheri (2005a, pp. 3–21).

that appropriate future modifications of our DNA, today not easy to foresee, could lead to new and more effective instruments and methods for nature investigation that could lead the future human being toward new, amazing discoveries on the origin and on the end of the universe as well as on everything that happens here, life included.

Let me add a final consideration on the bibliographic references used; all have been taken from my personal readings and from my life experiences, without any foreclosure on any points of views by any author, and without any (voluntary) ideological selection that could not have been consistent with my claim to avoid any emotional transport or with the search of a harmonic equilibrium between the two knowledge modalities.

Rosolino Buccheri
2020

Acknowledgments

I wish to thank Profs. Giovanni Boniolo and Francesco La Mantia for their illuminated foreword and afterword, respectively, as well as Claudio Vincenzo Piscopo for his contributed drawings.

I am indebted to my wife, Riri, for her continuous support and to my sons, Giuseppe and Mauro, for their care and contributions.

PART I
LAW, CHANCE, AND EVOLUTION

Enigma

From Nowhere's enigma, Universe springs;
From the night's black, Sun appears;
From non-existence darkness, Man arises,
In death's mystery later broken up.
In the night gloom, Sun vanishes;
Universe future in the arcane's hidden.

Amid start and closure, a lot we know;
To before and after we're not allowed.
Ever at present our conscience will halt?

—Rosolino Buccheri, 2014

Premise: Dissipative systems and self-organization processes

Before starting our discussion, I find it better to open an informative parenthesis, concise as it can be, on matter's self-organization processes in complex physical systems open to interaction with their environment.[1] In these systems, new structures may spontaneously emerge along their evolution path, with always more complex and unattended shapes with respect to the underlying fundamental physics, but anyway equipped with causal effectiveness in their following road.

Such a situation is basically due to the fact that in a complex system, the myriads of elementary particles of which it is made interact in a nonlinear way, reciprocally and with the environment, thus confusing any cause–effect relationship that is, instead, clearly visible in all linear processes. It follows that however precise the knowledge of the properties of single constituents and of their reciprocal interactive connections could be, any analysis of a reductionist type is usually ineffective, preventing us to foresee the role that every element will play in the future organization process of the system, therefore making us unable to explain any new emerging behavior.

Obviously, the difficulties in describing the evolution of any systems grow together with their level of complexity: from the simpler one, the universe evolution—mostly comprehensible by using physics and mathematics techniques—to the more complex one on living systems studied by biology, until the enormously more difficult study of the evolution of social systems, where only statistical simulation techniques may help. A hierarchy of complexity that, from the interaction of the elementary components of matter, leads to the evolution of the universe, then to life by means of the mutual interaction of complex molecules, and finally to man's knowledge with the emersion of the mind.[2]

---

[1]I refer to systems far from thermal equilibrium, those that Ilya Prigogine defined as "dissipative systems," among which we may include living systems, of particular interest for our discussion (Prigogine and Stengers, 1999).

[2]The first step toward living matter—the "informative property" and the forming of DNA—is seen in molecules with a very high level of complexity. The importance of this property, emerged along the evolution of matter toward life, will be discussed later.

Just following the opening by Ilya Prigogine of a new research stream on the non-equilibrium thermodynamics,[3] a deep investigation was undertaken about the above-discussed phenomena, where, countertrending with respect to the predictions of the second principle of thermodynamics, conditions may emerge for which their order may increase with time instead of decreasing.

By studying physical systems with a multitude of interacting elements in a state of thermodynamic disequilibrium, Prigogine realized that such systems gradually increase their complexity by means of nonlinear processes of self-organization, for which they acquire, almost temporarily, "ordered" energy and entropy, which is successively dissipated gradually into "degraded" energy and entropy. Prigogine discovered, in particular, that while in all complex physical systems in equilibrium, nature's laws are universal and predictable, when a system is far from equilibrium, it is submitted to changes due to the forces acting to it from the outside, and the laws describing it are not anymore universal and predictable but become specific of its present condition.[4] In such cases, any single physical system organizes itself through an effective and continuous exchange of energy and information with its environment, thus producing an order that is functional to its composition and to the features of the environment where it is merged so as to build and maintain stable for some time a well-defined internal structure. Along this process, we observe an evolution characterized by the continuous "emergence" of always new and unpredictable properties for which the quantum physicist Josef Maria Jauch wrote:

> "… the whole is more than the sum of its parts and […] the constructive integration of complementary processes is the secret of any activity in life."[5]

Here the term "more" refers to the new properties spontaneously and unpredictably emerged in the whole evolving system. A

---

[3]The non-equilibrium thermodynamics refers to non-isolated systems, open to interaction with outside. For such studies, the Nobel Prize for Chemistry was awarded to Prigogine in 1977.

[4]As Prigogine himself said, "Matter is blind when approaching equilibrium, when there is no more arrow of time; but when the arrow of time presents itself, matter starts to see!" (Prigogine, 1996).

[5]Jauch (2001).

circumstance, this, where we see the failure of the classical reductionism, for which the whole arrangement should be directly deductible from the properties of its single components.

The studies quoted before made Prigogine introduce in 1981 the notion of "dissipative structures" for this kind of systems and to think that their formation and their subsequent "chaotic" evolution by means of self-organization processes could be a general principle that all matter always obeys, in competition with the second principle of thermodynamics. A process in which the time evolution of every complex physical system develops through the formation of "chaotic" order, that is, through the tendency of inert matter to autonomously organize itself toward always more complex, temporary stable, structures. A trend that Prigogine assigned to the "creativity" of dissipation processes, able to effectively use the available energy.

Obviously, the debate on the "emergency" of new unpredictable phenomena in a physical system, starting from its elementary constituents, is far from being definitely closed. If from one side it seems already clear that any nontrivial emerging processes are bound to the specific nonlinearity characterizing the system itself, we may still ask whether it is possible to state a general theory of complex systems, able to indicate the necessary and sufficient conditions for the appearance of processes, irreducible to the laws that rule their elementary constituents.

Some elementary self-organization processes of matter in which spontaneous formation of order follows are known for a long time. One of the simplest, easily observable by everybody, is that of Bénard cells[6]—ordered configurations connected with the convective motion generated in a thin layer of liquid warmed from low.

A slightly more complex example, easily reproducible, is given by swinging chemical reactions, scientifically denominated "chemical clocks." The first of these changing structures, discovered by the soviet physicist Boris Belusov around 1950, was harshly challenged by the scientific community of that time and accepted only in 1964, when it was published by the biophysicist Anatol Zhabotinsky and thereby denominated "Belusov–Zhabotinsky reaction."[7] This process shows up in a mixture of potassium bromide, malonic acid, and manganese sulfate, prepared within a warm solution of sulfuric acid,

---

[6]Bènard (1901, pp. 62–144).
[7]Zhabotinsky (1964, pp. 306–311).

where the manganese oscillates between two different oxidation states, spontaneously causing a periodic variation of the mixture, lasting about four seconds. [8]

Besides Bénard cells and the Belousov–Zhabotinsky reaction, self-organization phenomena are involved in many other conditions of nonthermal equilibrium, typical of all dissipative systems. Classical examples are lasers and climatic or biological processes within living cells, such as enzyme and protein syntheses. Biochemical oscillations (biorhythms, in particular) are among the most important properties of the living systems and show up at all levels of the biological organization, from unicellular to many-cell systems, even with far different oscillation periods, from the fraction of a second to years.

Well known are, in particular, the results of studies done in 1953 by Stanley Lloyd Miller, guided by Harold Urey (Nobel Prize for Chemistry, 1934). They referred to the observed spontaneous formation of basic-for-life molecules (actually a small quantity of two amino acids, alanine and glycine) a week after having submitted a mixture of water, ammonia, and methane (afterword called "primordial soup" by Alexander Oparin and John Haldane) to electric discharge.

These experiments were repeated in 2002 by NASA astronomers, who succeeded in producing alanine, glycine, and serine after irradiation of some low-temperature water mixed with simple molecules (afterword denominated "spatial ice," apt to simulate the typical conditions of interstellar space) with ultraviolet light.

These results confute the hard criticism always addressed to all kind of studies interested in investigating the possibility of the spontaneous formation of life. Actually, even by considering that the primordial soup could not precisely represent the composition of the original terrestrial atmosphere, which is assumed, instead, to be rich in nitrogen, carbon dioxide, and water vapor,[9] and by admitting that within such a mixture the molecules, including carbon atoms, could not have a long life, the spontaneous formation of amino acids from its elementary components remains an established fact to be taken heavily into account. In addition to this, there are the results

---

[8]See Buccheri (2008, pp. 85–96) for a deeper discussion and related figures.

[9]A more detailed analysis of the state of knowledge about the emergence of biomolecules from which life may develop is proposed by Boniolo (2003).

of other researchers, revealing the presence of the elementary constituents of life from other regions of the solar system, where the physical conditions may be different from those existing on Earth[10]— results affirming the presence of amino acids in some meteorites,[11] thus demonstrating that organic molecules may be everywhere in the universe, probably originated by synthesis processes of heavy elements during the formation of stars, beginning from hydrogen, the lighter element.

Such facts lead us to a hardly escapable conclusion, that is, the existence of abiogenesis processes that, starting from inorganic chemical stuff, may generate organic chemical compounds. A starting event, this one, of the complex transition from the inanimate world to living matter.

On the basis of these data, it looks well possible that the chain of events leading to life could have started in the interplanetary space where the chemical elements produced by fusion in dying stars could transform, having available sufficient time, into organic molecules that, by finding adequate physical-climatic conditions (e.g., in a planet), may grow in complexity until being capable of metabolism and reproduction.

Although the path to acknowledge definite affirmations satisfying the requested scientific rigor in all their details is still long and full of difficulties, on the basis of all that is expressed before and taking into account the not easily questionable studies on Jean-Baptiste de Lamark's and Charles Darwin's evolution of species, I don't believe it strange to look at living systems as very complex physical systems derived from the tendency of matter to spontaneously self-organize, starting from its most elementary components. Systems that, just because they evolve together—as well as in competition—with the environment, once formed, are able to run their own vital process far from the thermodynamic equilibrium,[12] getting from outside the energy necessary to their maintenance and to the production of their

---

[10]To be cited, in particular, the *panspermia* theory with all its variants, due to Francis Crick and Leslie Orgel, to Swante Arrhenius, to Juan Orò, and to Fred Hoyle and Chandra Wickramasinghe.

[11]Martins et al. (2008).

[12]Such activities are ensured in human beings by periodic processes able to supervise cardiac and respiratory functions such as the sleep–waking circadian rhythms that have notoriously a fundamental role in life stability.

emerging properties, useful to increase their level of complexity so as to keep them in good, although temporary, stability along their evolutionary path.

Along this line of thought, the biologists Humberto Maturana and Francisco Varela developed (1992, 1993)—just in view of the specific complexity of living organisms—the concept of an autopoietic system,[13] that is, a net structure with its nodes in *feedback*, reciprocally and with the external environment. A system able to produce ex novo its own components, in some cases able to eliminate or modify their functions, with the aim of maintaining the system stable for as long as possible.[14]

---

[13]Maturana and Varela (1992).
[14]Maturana and Varela (1993).

Chapter 1

The Evolution of the Universe

The beliefs of the seventeenth's century, for which the world would have been created about 5000 years before,[15] started to be questioned in cosmology with the *Theory of the Primitive Nebula*, originally published in 1755 by Immanuel Kant and more rigorously resumed by Simon de La Place in 1795. La Place supported the idea that a primordial cloud of particles progressively condensed under the action of the gravity attraction, thus giving origin to the sun and to planets. Such a general scenario is considered possible still today, even with more precise details, consisting, for example, in the fact that the primordial cloud could be composed by hydrogen, helium, and powders of heavier elements. La Place assumed, in addition, that such a cloud would have suffered, during millions of years, the pressure of a shockwave consequential to the explosion of a nearby supernova. The result would have been a disc of very hot material from which a *protostar* would have escaped—lately converted to a star—while its fragments, expelled by the centrifugal force, would have formed, by means of continuous reciprocal collisions, planets with their satellites. From these original ideas on the birth of the solar system and from the observations of Edwin Hubble about galaxy recession, in the middle of the last century, the concept of evolution of the universe was established.

---

[15]Johannes Kepler fixed in 3877 BC the starting date of the universe; the archbishop James Ussher brought it forward to 4004 BC, while Johannes Hevelius proposed the year 3963 BC.

Myth, Chaos, and Certainty: Notes on Cosmos, Life, and Knowledge
Rosolino Buccheri
Copyright © 2021 Jenny Stanford Publishing Pte. Ltd.
ISBN 978-981-4877-33-6 (Hardcover), 978-1-003-08869-1 (eBook)
www.jennystanford.com

1.1 From Lemaître's Prediction to Hubble's Observations

In 1927, Georges Lemaître, cosmologist and theologian, published *The Hypothesis of the Primordial Atom*, in which he conjectured an expansion law from which the universe would have been originated. Two years after Lemaître, Hubble, measuring the motion of some galaxies by means of the *Doppler effect*, hypothesized the reciprocal moving off between galaxies. These results, not firmly established at that time, were first opposed by Albert Einstein[16] and successively also by Fred Hoyle, Hermann Bondi, and Thomas Gold, who, in 1948, proposed the *theory of stationary state*, consonant with the *fissism* of living species.[17] Later *Doppler* measurements, using the discovery of many other galaxies, confirmed the reciprocal galaxies' moving off, which allowed to establish the relationship $v = H_0 D$ between the recession velocity v of galaxies and their distance D from Earth, definitely establishing the expansion of the universe. The *theory of the stationary state*[18] was set aside and the new concept of *evolution* was promoted in cosmology.

In the middle of such a flurry of ideas, in 1938, Pierre Teilhard de Chardin, biologist and theologian, published *Le phénomène humaine*, where he hypothesized the evolution of the universe from a beginning—defined by Lemaître as the α-point—up to a final ω-point, passing through successive phases that, one after the other, came up to reality: the inorganic world, the primitive life, the advent of man, and the appearance of his conscience. A hypothesis, this, that triggered the negative reaction of the Church, which accused Teilhard of pantheism.

---

[16]In opposition to Lemaître's hypothesis that could let us think to creation, Einstein modified the general relativity equations by introducing a "cosmological constant" in order to obtain a static universe. Following the confirmation of the galaxies' recession, proposed by Hubble, Einstein declared that the cosmological constant was the greatest mistake of his life (see, for example, Fischer, 1995).

[17]The *fissism*, supported by Linnaeus and others in the nineteenth century, had taken up Aristoteles' ideas, by which, in the absence of any hypothesis on their origin, all living systems existed since ever, exactly as they were actually seen.

[18]The theory of the stationary state postulated a static universe, without start nor end. To obey Hubble's law $v = H_0 D$, it had to assert the existence of a constant density through a continuous creation of matter.

1.2   From the Big Bang to Particles

Today's cosmology tells us that the universe was born about 14 billions years ago by explosively emerging from a bubble of quantum void from which space, time, and matter were originated, as we see them today. An explosion described by the physicist Guido Tonelli as

> ". . . just one bubble, similar to others, but that instead of immediately closing down and returning to the fundamental state [. . .] in an extremely small time, it expanded at an incredible velocity and assumed enormous dimensions."[19]

and by the physicist Fernando De Felice as a "symmetry breaking" of the "primordial broth" caused by an occasional fluctuation, amplified by the presence of a negative pressure already existing on the initial magma.[20] From this event—variously described and sarcastically called *Big Bang* by Fred Hoyle—all elementary particles (electrons, neutrinos, quarks . . .) and fundamental forces (gravity, electromagnetic, nuclear . . .) arose. From their combined action, a mass of atoms, composed of three quarters hydrogen and one quarter helium, emerged just after a few minutes, together with some slightly heavier elements. Both, elementary particles and fundamental forces, are characterized by numerical constants, called "nature's constants" (atomic energetic levels, chemical bonds, fine structure constant . . .), because no variation has ever been observed in their values.

1.3   From Thermonuclear Fusion to Neutron Stars

The Big Bang and the subsequent expansion of the universe have caused, during the course of time, a series of extraordinary phenomena occurring under gravity's action, in strict collaboration with all other force fields. After a few hundreds of millions of years from the Big Bang, various small density fluctuations (statistically present everywhere) started to attract by gravity the matter

---

[19]Tonelli (2017, pp. 44–45).
[20]De Felice (2012, pp. 20–27).

around,[21] causing their agglomeration in gigantic, very cold clouds, within which *protostars*[22] started to form. As long as the mass of such a *protostar* became heavy enough, the gravity would increase its contraction, thus causing further growth, both in temperature and in internal density, until when the short-range nuclear forces, now prevailing over gravity, would determine the trigger of the thermonuclear fusion reaction and the conversion of the *protostar* into a bright *star*. If, anyway, the mass of the agglomeration would be too small, the mechanism of fusion from hydrogen to helium would not be so effective and the *protostar* would slightly transform itself into a *black dwarf* that would shine for billions of years at low luminosity. Still smaller agglomerates of matter may form objects like Jupiter or Saturn, planets that can well be considered unfinished stars. Moreover, the dust and debris produced from the original cloud along the whole process of formation of *protostars* and *stars* gathered slightly together, thus giving rise to a disc of rotating matter whose turbulent motion would induce continuous reciprocal collisions, which, in turn, might have caused both further crushing and further agglomeration toward always larger rocky systems until the formation of planets of various dimensions.

During the thermonuclear fusion process, stars are kept in equilibrium by a cyclic expansion–contraction mechanism for which the star reacts with expansion to every increase of energy produced by fusion, and the expansion, in turn, causes a decrease of pressure and temperature, triggering star contraction by gravity, thus producing fusion energy apt to let it expand again. In the course of this cyclic process, high-energy impacts between hydrogen and helium atoms, initially in equilibrium, cause them to fuse together, forming heavier elements: First, the hydrogen transforms itself into helium, but as soon as the hydrogen becomes insufficient for effective production of helium, the energy produced by the star starts again to decrease. At this point, further evolution depends on its total mass: if it is almost equal to that of our Sun, the fusion process from hydrogen to helium does not proceed further; the production of energy, as already said, decreases and the star contracts. Its

---

[21]The first condensation interested basically hydrogen and helium; lately the space will be enriched by heavier elements coming from supernova explosions.
[22]Protostars are single lumps of matter stabilized in temperature and density by pressure due to the motion of the particles.

internal nucleus becomes more compact and its radius, at the same time, temporarily increases, as does the temperature of its external shell, so transforming the star into a *red giant*.[23] Afterword, the lack of fusion alimenting the internal fuel brings back the temperature to lower levels, such to further transform the *red giant* into a *white dwarf*, a little star, as large as our Earth, where the force of gravity is balanced by the pressure of the *degenerate gas*—as expressed by Pauli's principle[24]—that slightly cools down, finally ending in a not too much luminous *black dwarf*.

If, instead, the total mass of the star is at least one and a half times larger than that of the Sun, the impacts increase together with the internal star temperature, thus causing fusion into elements heavier than helium. Successively, for even larger star masses, lithium, beryllium, boron, carbon (the element constituting living beings), nitrogen, oxygen, and so on until iron are generated.

The fusion of iron, anyway, does not produce energy but absorbs it, thus inducing gravity to take over the radiation pressure. An event, this, that implies the *supernova* collapse, consisting in the star's implosion over itself, with the simultaneous emission out in the space of an enormous quantity of energy and atomic material so as to equal, even if for only a limited period, the luminosity of an entire galaxy. A catastrophic event, this, in which a multitude of neutrinos, produced immediately before the collapse, interacts with the superficial gaseous layers of the star, already rich in heavier elements, pushing them at very high velocity toward interstellar space[25] with the consequence of further compressing the particles of the external cloud. The enormous pressure caused by this impact triggers the formation—within the cloud itself or nearby clouds—of new stars (then called "population 1 stars") with a chemical composition enriched in heavier elements with respect to those formed with the original gas, containing only hydrogen and helium.

Concerning the original exploded star, it remains only an enormous, very dense residual nucleus of matter whose internal

---

[23]Our sun finds itself in the middle of this phase, called the "main sequence." It will still last 4–5 billion years.

[24]Pauli's exclusion principle states that two identical fermions (protons, neutrons, and electrons) cannot simultaneously occupy the same quantum state, stating a limit to the contraction of the ordinary matter.

[25]This phenomenon is called "stellar wind."

gravity attraction is so great as to win even the resistance opposed by Pauli's principle and therefore to reduce its dimensions until a diameter of only 10 km! An object thereby called *neutron star* made up by practically only neutrons, dizzily rotating and emitting two powerful electromagnetic radiation beams from its two poles.

1.4   Pulsars, Black Holes, and the Future of the Universe

Concerning the residual neutron star, it may happen that the axis of rotation movement is oriented toward Earth. In this case, we see the emitted radiation to pulsate synchronically with the rotation period, and the neutron star is called "pulsar." Sometimes we see pulsars orbiting around other celestial objects (binary pulsars), often a white dwarf or another neutron star, or even a system of two pulsars orbiting one around the other. These last objects are extremely important for physics, since the very intense reciprocal gravitational field allows us to verify Einstein's theory of relativity, as it happened in the case of the binary pulsar PSR1913+16, discovered in 1974, and of others known since then. It has to be noted that while single pulsars gradually decrease their rotation velocity because of the emitted radiation, binary pulsars, by sucking matter from the companion star, rotate always faster until reaching a very high speed so as to complete one tour in a very short time, even only one-thousandth of a second![26]

Finally, if the mass of the residual nucleus is much greater than that of our Sun, a *black hole* may be generated. An object, this, whose enormous density may create around it such large a gravitational field that even time may be substantially deformed, not only space, braking down the static Newtonian linearity. A deformation, the last, able to deviate everything around the *black hole*, even the path of light that could be confined within a region called *event horizon*.

All the diverse dynamic configurations of astronomical objects quoted in this chapter, together with others not quoted here, and

---

[26]The first of these superfast binary pulsars was discovered in 1983 by the writer in collaboration with Valentin Boriakoff and Franco Fauci at the large Arecibo radio telescope in Puerto Rico (Boriakoff et al., 1983).

still others that are continuously being discovered by means of always more powerful telescopes, both from Earth and from spaceships, although appearing to us united in their fascinating and phantasmagoric variety, are instead precisely ruled by the four fundamental forces and bound to the starting eject power due to the initial Big Bang. By means of such a powerful impulse, the universe is still expanding and cooling down with the result that if this situation goes on as we observe it today, in a far future all the matter of the universe could be concentrated in a myriad of black holes dispersed in an infinite cosmos until, as Tonelli writes,

> "there will no more exist sufficient energy for any further evolution, and everything will end with a shroud of dark and cold that will wrap a necropolis of stars."[27]

By commenting on the possible consequences of the precarious stability of the electroweak vacuum, just budgeted from the recent discovery of the long-time-searched Higgs boson, Tonelli hurries to clarify that whatever it will be, the universe should not necessarily end in this way, since for such a precariousness

> "already at its born, just a bit would have been enough to let become everything totally instable: a Higgs boson slightly lighter and the microscopic tearing of the primordial vacuum, opened few instants before, would immediately close and all would end before beginning [...] that slim scaffold would have crashed under the pushing of one of those scary catastrophes that sometimes interest the farthest galaxies [...] the entire universe would have vanished in an immense bubble of pure energy."[28]

Let me add here, probably for fun, the possibility of a *Big Crunch*, that is, the inverse of the Big Bang, the return singularity that could be materialized in the case in which the *dark matter* and the *dark energy* would be greater than today imagined, thus determining the inversion of the process of expansion toward the reduction of all the matter of the universe in a compact nucleus of matter and energy, analogous to that pre-existing the Big Bang.

---

[27]Tonelli (2017, p. 155).
[28]Idem, pp. 157–158.

Chapter 2

The Origin of Life and Its Evolution on Earth

Let me propose, here, a brief description of the origin of life on Earth, based on today's scientific fonts, with special reference to the vast class of conditions necessary for its emergence and further development. This description offers the idea that, as it happens for the whole universe, life itself may be in continuous evolution in all its aspects, which makes us venture on a possible hypothesis upon which to base a "sufficient" condition for the actual concretization of the vital process.

The information derived from the geologic history of Earth and of our solar system, together with the data collected about dissipative systems, gives us good reasons to think that the emergence of live could have been the result of a chain of self-organization processes of elementary matter, in *feedback* with the environment. A view, this, contrasting the ancient belief of an ad hoc intervention from a god, external to the universe, aimed at the creation of all living systems as we know them today.

In 1809, the naturalist Lamark published *Philosophie Zoologique*, in which he suggested that all modifications to living systems produced by their milieu might be transmitted to successive generations. Following Lamark, Darwin, in 1859, by opening his *On the Origin of Species by Means of Natural Selection* with a *Historic*

Myth, Chaos, and Certainty: Notes on Cosmos, Life, and Knowledge
Rosolino Buccheri
Copyright © 2021 Jenny Stanford Publishing Pte. Ltd.
ISBN 978-981-4877-33-6 (Hardcover), 978-1-003-08869-1 (eBook)
www.jennystanford.com

Compendium of the Progress of Ideas on the Origin of Species, wrote with reference to fissism and creationism that

> "Until not so long, the great majority of naturalists thought that all species would be immutable and would have been created one independently from the other."[29]

    With Darwin's publication, the concept of evolution was definitely established in biology; according to it, the origins and development of living organisms in our universe are seen as the product of a series of processes governed by natural laws in a temporal succession of events characterized by irreversibility and unpredictability.[30] A circumstance, this, for which Francisco Ayala (2005) writes,

> "The theory of evolution manifests chance and necessity jointly implicated in the problem of life; case and determinism twisted together in a natural process able to elaborate the most beautiful and diverse entities of the universe."[31]

2.1   The Anthropic Principle

This principle, introduced in 1973 by Brandon Carter[32] and successively resumed by John Barrow and Frank Tipler (1986), proposes that the existence of human life implies stringent constraints on the configuration of the universe as it came out of the Big Bang and on its subsequent evolution. In its weak variant (*weak anthropic principle* [WAP]) Carter specified that the values of nature's constants[33] must be compatible with the existence of

---

[29]Darwin (1967, p. 67).

[30]For our discussion it is not important whether the evolution is gradual, *à la Darwin*, or is discontinuous, as stated by Stephen Jay Gould in his *Punctuated Equilibrium* (Gould, 2007).

[31]Ayala (2005, p. 30).

[32]The term "anthropic principle" was coined by Brandon Carter during the IAU symposium held in Krakow (1963) on *Comparison of Cosmological Theories with Observational Data* (see Carter, 1974). This symposium was organized to celebrate the 50th anniversary of Copernicus's birth.

[33]"Nature's constants" are the numerical values of those physical quantities considered independent both from time and from the place in which they are measured. Examples are Planck's constant $\hbar$, the gravitational constant G, the light velocity in vacuum c, the dielectric constant of vacuum ε_0, the electric elementary charge e, masses of protons and electrons, the fine structure constant, and many others.

human observers.[34] A consideration, this, that became in Barrow and Tipler's (1986) version

"The observed values of all physical and cosmological quantities are not equally probable but they take on values restricted by the requirement that there exist sites where carbon-based life can evolve and by the requirement that the Universe be old enough for it to have already done so."[35]

In its "strong" variant (*strong anthropic principle* [SAP]), always according to Barrow and Tipler (1986), the principle establishes that

"The universe must have those properties which allow life to develop within it at some stage in its history."[36]

Furthermore, in 1986, Barrow and Tipler, by emphasizing the informative properties intrinsic to matter, proposed a third version, the *final anthropic principle* (FAP), according to which

"Intelligent information-processing must come into existence in the universe and, once it comes into existence, it will never die out."[37]

While WAP and SAP constitute, de facto, the necessary condition from the development of life, FAP with its "must," would have been a sort of sufficient condition, even if of axiomatic nature, without a clear justification. By recalling the "informative" property stated by FAP and emerged at a certain level of complexity, Manfred Eigen writes that

"Any information arises from a lack of information. Furthermore, it is not simply an information making visible a pre-existing information: actually, once the information is generated, the state of the system appears of an absolutely new type."[38]

A property, therefore, that, once emerged, governs and manages—as foreseen by FAP—all biologic processes, beyond

---

[34]Idem, pp. 291–298.
[35]Barrow and Tipler (1986, p. 16).
[36]Idem, p. 21.
[37]Idem, p. 23.
[38]Eigen (1992, p. 51).

the basic physics and waiting for that further, still more powerful, quality jump, still more mysterious from the physical point of view, that in the future progress of evolution, at an even higher level of complexity, announced the emergence of mind and conscience.

Carter's attempt, as well as that by Barrow and Tipler, was to highlight the compatibility of biological life with the laws ruling the universe and therefore suggest caution to scientists when divulging observational data in cosmology before having first excluded any incoherent interpretative consequences.

The criticalities of nature laws will be discussed later. It is here enough to quote that the anthropic principle has been always variously commented and interpreted and sometimes used in order to justify totally different points of view. Giuseppe Tanzella-Nitti, in particular, points at the anthropic principle as a revaluation of the central position of the human being in the universe after his downgrading in the sixteenth century with the Copernican principle, thus attributing to it a finalistic value.[39]

2.2 Nature's Constants and the Conditions for the Existence of Life

The inventory of the conditions needed for the existence of life on Earth is very long, and besides physical ones, it includes also all those constraints connected to the fine equilibrium of the already cited nature's constants, small variations of which could have given origin to a universe with a configuration and a set of natural laws dramatically different from the present ones and, anyway, not compatible with the emergence of life. Let us resume here some particulars of such criticalities, starting from the most general ones.

The first concerns the relationship among the various forces. Out of the four fundamental ones, gravity ($F = Gm_1m_2/r^2$) is by far the weakest (10^{39} times smaller than the "strong interaction")[40] and with no relevant effect on subatomic particles (electrons, protons,

---

[39]Tanzella-Nitti (2002).

[40]The "strong interaction" (or strong nuclear force) acts within atomic nuclei, where it keeps together neutrons and protons. At an even lower scale, it keeps together quarks to form protons and neutrons. It is about 10^{25} times weaker than the "weak interaction" but only 100 times weaker than the electromagnetic force, and together with this last, it constitutes the "electroweak interaction."

neutrons, quarks); conversely it has a range of action almost infinite, allowing it to affect every celestial body and the expansion rate of the universe.

If the ratio between gravity and other forces would have been different as it actually is—even a little bit less or more—the universe could have emerged from the Big Bang with a content of hydrogen much lower than the ¾ known, with the consequent impossibility to have water or other compounds of hydrogen. Vice versa, if it would have emerged with an insufficient content of cosmological helium, the consequence would have been that the evolution cycle of stars would have drastically changed, thus negatively modifying the transformation activity of all elements.[41]

The gravitational law contains, too, an important criticality because of its dependence on the square of the distance, which is essential for the existence of the solar system. A dependence different from the square of the distance could not have allowed planets to have stable periodic orbits but spiraling trajectories finally falling to the Sun or escaping away toward space. Furthermore, if the gravitational constant G would have had a value slightly larger than the measured one, the stars would have become too warm in too little time and the universe would have collapsed within itself, avoiding any further evolution. If, on the other side, G would have been slightly smaller, the formation of galaxies and stars (and, a fortiori, of planets that originated along the formation process of stars) would have never started, and life, as we know it today, would have never been born.[42]

Earth, and ourselves, being composed of elements heavier than hydrogen, are an integral part of such a continuous transformation cycle of matter and energy that takes place by means of nuclear fusion within stars. Living matter, in particular, is formed of about 83% carbon and oxygen and 13% nitrogen, and the carbon has the very important characteristic to be able to form long molecules in combination with other atoms, even with multiple bonds; quality, this last, that allows carbon to bind with hydrogen, oxygen, and nitrogen so as to form the extraordinary complex molecules forming all living organisms: fatty acids and glycerol (forming

---

[41]Barrow (2004).
[42]Ibid.

lipids), monosaccharides (carbohydrates), amino acids (from which proteins are born), and nucleotides (forming RNA and DNA).

Still concerning criticality, the fundamental production process of carbon formation within stars is not less important. As we see from the entire sequence of events characterizing its production, the process of carbon formation shows a surprisingly delicate equilibrium. First of all, three helium nuclei (α-particles) must converge in a point and interact each other; furthermore, to have a significant quantity of such an essential element, there must be a fast and effective process of overproduction in such a way that the next transformation into oxygen should not destroy the carbon at all. To obtain such an overproduction, a process of "resonance" occurs consisting in the fact that the sum of the energies of the reacting α-particles is very close to the natural energetic level of the just-forming carbon atom. If we look into the details of this process, we observe that at the beginning two helium nuclei combine together, giving out a beryllium nucleus, which, by having a very long average lifetime (10,000 times more than the interaction time between two helium nuclei), stays around so long as to get an opportunity to be combined with a third helium nucleus, thereby forming a carbon nucleus. The energetic level of 7.656 MeV of the carbon nucleus stays *just* above the sum of the energies of beryllium (only 0.0011 MeV higher) and of the third helium nucleus, a circumstance that makes the reaction become "resonant" by means of the little energy of the thermic movement within the star, thus allowing the formation of large quantities of carbon. De facto, very little beryllium is formed, but a much larger quantity of carbon is saved, more important than beryllium for biologic processes.

The subsequent reaction leading to oxygen through the capture of helium ($C_{12} + He_4 = O_{16} + \gamma$) is also extremely critical. If the energetic level of oxygen would not be just a little bit smaller to that needed for the reaction, carbon would be completely burnt to produce oxygen.

To escape all these fortunate circumstances, so avoiding to trigger discussion of religious type, Leonard Susskind, in his book *The Cosmic Landscape. String Theory and the Illusion of Intelligent Design*, imagines a multiverse, a set of myriads of separate universes, each one with a different set of nature's constants and only one where we are able to live. Other, more concrete scientists, hypothesize that the values of nature's constants may vary within the universe so as to allow life only within those regions where they assume values compatible with life. The real problem, perhaps, is that we will never know in all its details our universe—in its present and in its future— or even know universes others than this one. These hypotheses could then remain beautiful theoretic constructions forever, without any verification possibility. Before retreating ourselves to fantasy, therefore, it is maybe better to look as deeply as possible into our present, although limited, knowledge.

2.3   The Physical-Climatic Conditions on Earth

Earth is the only place presently known with ideal physical and chemical conditions for the birth and evolution of life—then called anthropic conditions—where it is active with a multitude of very complex living beings. The constant presence of water in liquid state[43] plays a fundamental role, together with the favorable distance from the sun that, by allowing a range of temperatures placed between freezing and boiling points, reduces the water's evaporation for excessive heat and its hardening for excessive cold. Other important conditions for maintaining life are the presence of an atmosphere (for its role in alleviating the temperature's excursions between night and day and between seasons) together with the presence of a magnetic field screening the deadly solar radiations and of the moon that, with its reasonable gravity attraction (not too large mass and distance from earth), stabilizes the 23° inclination of Earth's axis with respect to the orbit, thus allowing a regular alternating of seasons. To be noted, finally, is also the protection got from the giant planets Jupiter and Saturn that, due to their more external position in the solar system and their large gravitational attraction, deviate on themselves the multitude of asteroids and comets coming from outside the solar system, avoiding devastating impacts on Earth's surface.

2.4   The Epochs of Life's Evolution on Earth

Under the just-discussed cosmologic, physical, and climatic conditions, the development of life until the appearance of the

---

[43]So maintained because of the reasonable distance between Earth and the sun.

human species, in the more than four billion years' history of Earth, has occurred, following Christian de Duve, through seven epochs, corresponding to seven levels of complexity: the epochs of chemistry, of information, of the protocell, of the single cell, of the many-cell organisms, of the mind, and, finally, of the unknown, these last two not considered here.[44]

The protagonists of the epoch of chemistry are the elements derived from the previous nuclear fusion processes, produced by deterministic laws, able to favor the formation of chemical elements basic for life, and their reciprocal interaction—carbon, oxygen, hydrogen, and nitrogen that, thanks to supernova explosions, are distributed everywhere in the universe. These elements combine together in order to form, first, inorganic molecules like water (H_2O), ammonia (NH_3), and carbon dioxide (CO_2), and simple organic molecules like methane (CH_4). These molecules, finding themselves in an environment where water and other chemical substances are present in a sufficient quantity, bind together little by little, ending to form with time always longer and more complex biomolecules like glycerol, amino acids, and monosaccharides.

The following epoch, that of *information*, shows a quality jump with respect to the elementary laws of physics and chemistry, because it registers the appearance of new configurations, still not completely understood but anyway able to develop matter until such a degree of complexity as to succeed in giving origin to nucleic acids, to proteins, and to other complex molecules, basics of life, and still lately, to the formation of the double-helix structure of DNA, up to 2 m long, discovered in 1953 by James Watson and Francis Crick.

With the formation of DNA, just before the appearance of the living cell, unexpectedly emerges a "project"[45] aimed to govern hereditary continuity through the replication of genetic messages (thereby transmitted to further generations), which, by means of sudden and unpredictable mutations, managed by the natural selection (defined by Francisco Ayala as "opportunistic process"[46]), connotes the future evolution. For this purpose, Christian de Duve (1998) writes

---

[44]de Duve (1998, p. 10).

[45]The term "project" means the ability of the property "information," acquired by inert matter, to "plan" the development of the living being features.

[46]Ayala (2005, p. 37).

"mutations are accidental events that, according to some people, imply the view of an evolution dominated by chance. While not denying the role of contingency in evolution, I underline that chance is anyway pulled by certain constraints—physical, chemical, biologic, environmental—limiting its free path."[47]

The property to be equipped with a project (the *telenomy*) is, according to Jacques Monod (1986), the necessary condition to define a living system, whereas the sufficient condition is to be looked for (always according to Monod) in the author of the project, that is, in the forces external to it, or to the ability of the author to reproduce himself.[48] Once capable of replication, the biomolecules organize themselves within the protocell, which can therefore be considered the common progenitor of all forms of life present on Earth. A structure *á la Prigogine*, provided of a membrane act to protect itself from the external environment and to allow nutritional substances to enter and wastes to exit.

From the protocell, born about 3.8 billion years ago (soon after the formation of the solar system), the epoch of the single cell followed, with the evolution and diversification of prokaryotes, bacteria able to use the energy of the solar light in order to decompose water into its components, hydrogen and oxygen. Between two and one and a half billion years ago, oxygen radically modified Earth's atmosphere, becoming the most important element for human life. Successively, to adapt themselves to the new environment, already oxygen-saturated, the prokaryotes transformed themselves, through symbiotic alliances occurring along a billion years, into eukaryotes, thus producing the unicellular organisms that today contribute to forming all living systems.

With the following transition to many-cell organisms, Earth started to be inhabited by a great variety of plants and animals with increasing complexity (before in water and later on land) and variety produced by casual mutations, indicated by the term "biodiversity." Within this context, following the evolution of the

---

[47]de Duve (1998, p. 12),

[48]Monod (1986, p. 22). The formulation of a "sufficient condition" expressed by Monod refers to concepts as "external forces" or "author of the project." We shall discuss it later.

brain and of reproductive strategies,[49] the common progenitor to all anthropomorphic species existing today came out six or seven million years ago. The human being descends from this progenitor. As Francisco Ayala (2005) writes, the human being

"[. . .] is a biologic species evolved from other non-human species [. . .] Our closest biologic relatives are the big monkeys,"[50]

among which, according to Boniolo (2003),

"about 27 million years ago we may identify the one called *proconsul*, from which the super family—Hominoidae—to which we belong, derives. About 24 million years ago, also the family Hylobatidae— gibbons and siamangs—came out, as well as the orangutan's family Pongidae, 20 million years ago. Finally, 14 million years ago was the turn of our family, Hominidae that, about 10 million years ago would differentiate in the subfamily Paninae, to-day including gorillas, chimpanzees and bonobos, and in the other subfamily Hominae."[51]

The last species evolved until the appearance in Africa of the first hominin, the *Ardipithecus ramidus*, perhaps biped, which lived 4.4 million years ago, possibly our founder. The *Australopithecus anamensis, A. afarensis, Homo habilis, H. erectus, H. ergaster* (so called "Turkana's boy"), and *H. neandertaliensis* followed to the *Ardipithecus ramidus*. Finally, the *H. sapiens* came, who, from 130,000 to 160,000 years ago, in the Paleolithic age, invaded Europe with a creative explosion testified by very impressive artistic, musical, and symbolic activities for which Ian Tattersall (2004) wrote that the art of the Cro-Magnon[52] was

"much more than a mechanic interpretation of the surrounding environment [. . .] It was, instead a complex recreation of the external

---

[49]I refer to the recent news indicating the discovery—made by a collaboration of Belgian and American scientists—of the gene *NOTCH2* that 3.5 million years ago started to differentiate until giving origin to three new genes. Their activity has increased three times the number of the neurons in our brain with respect to that of the *Australopithecus*.

[50]Ayala (2005, p. 57).

[51]Boniolo (2003, p. 47).

[52]The "Cro-Magnon," of the species *Homo sapiens*, lived almost 40,000 years ago. The name was derived from the French region *Abri de Cro-Magnon*, where the first fossil was found in 1868.

world of which we will never know with certainty the mythical context
[…] but it is clear the symbolic meaning contained in the superb images
of the animals living together with them, a meaning transcending the
simple zoological identity."[53]

Such witnesses let us think that man's symbolic ability must
have occurred much before the start of the official history, put by
archeologists and anthropologists between 4300 and 3000 BC, the
period where mythology places the end of the happy "golden age."[54]

2.5 Looking for "Sufficient" Conditions

The previous discussion allows us to interpret the origin of human
beings as the result of the development of life, starting from initial
organic elements. Elements that, anyway, have been generated on
Earth or brought there together with water, by means of impacts
of meteorites or asteroids rambling within the solar system. By
assuming this interpretation as correct, it remains, however, to be
understood how and why organic compounds may bring to life,
starting from inanimate matter (on Earth or in other parts of the
universe), in view of the evidence that the myriads of species of
microorganisms, plants, and animals—without excluding ourselves,
able to investigate nature and life—are made up of exactly the
same ingredients of which also the inert matter is made. How was
the process by which inert matter transformed into living matter
triggered, along the self-organization process, starting from the
nuclear fusion within stars?

Leaving out of consideration the mythic idea of a creation ex
nihilo—that still today, together with the belief of a flat Earth,
is shared by a number of unaware people—the process of the
formation of biotic molecules was thought, until 40 years ago, to be
simply the result of chance combinations among atoms. A chance

---

[53]Tattersall (2004, p. 160).
[54]Around the year 4000 BC, the celestial path of the Milky Way started from the spring
equinoctial point, located in the Constellation of the Twins, and ended in the autumn
equinoctial point, located in the Constellation of the Sagittarius: two points believed
by humans to be ports from where to enter into the sky. The equinoctial precession
movement deranged such a configuration, so disconnecting the Milky Way from these
two ports as seen from Earth and decreeing the end of the "golden age." See Bocchi
and Ceruti (2006, p. 21).

that, in a long course of time, realized the way to self-replicate, so casually producing the living cell. Such an explanation, mainly based on statistical considerations, was debated for a long time, at least since when Frank Drake, in 1961, proposed an equation aimed to estimate the probability of existence of intelligent life in the universe.[55] An estimate, actually, too approximate and based on too many hypothetic numbers to have a reasonable validity. From his side, George Wald (1954), Harvard University, estimated as three billion years the time necessary for nature to find, among all possible casual combinations, the correct one and, by considering the enormous number of Earth-like planets existing in the cosmos, concluded that life could well have been born spontaneously on our planet. He wrote

"By giving to it enough time, the impossible becomes possible, the possible probable, the probable, virtually certain. We have only to wait: time will do the miracle."[56]

Twenty years after Wald, a colleague of his at Harvard University, Elso Sterrenberg Barghoorn, discovered, by means of a scanning electronic microscope, few fossils of bacteria perfectly developed within rocks whose age was estimated to be 3.8 billion years, practically simultaneous to the same formation of Earth if we take into account the time necessary to accumulate a sufficient layer of liquid water at the needed temperature. Wald's statistical motivation, at least for what concerns Earth, could then falter but Sterrenberg's calculations were contested at the cosmic level by Christian de Duve. He, in his *Vital Dust*, wrote that the human body contains between 75 to 100 billion cells that never have simultaneously the same action but that intervene, each with its own contribution in strict coordination but independently from each other, with the aim to produce the wonderful life activity known, thus skeptically commenting on the probability of the spontaneous explosion of life. In his words, according to de Duve,

---

[55]Let us remember that Drake's formula aimed to roughly estimate the number of extraterrestrial civilizations possibly present within our galaxy on the base of other, very approximate, estimates: the presumed number of planetary systems with planets able to host life and the average rate of formation of planets where life could develop until the birth of intelligent beings capable of communicating.
[56]Wald (1954, p. 50).

"Life could be emerged out of a fantastic stroke of luck [. . .] Even if billions of planets could have had an evolution similar to that of the Earth, and even if billions of *Big Bangs* could have given rise to universes like our one, life could have never appeared. Its origin has been a *luxus naturae*, a cosmic joke,"[57]

so agreeing with Monod when he writes that

"[. . .] the universe was not giving birth to life nor the biosphere was giving birth to man. Our number got out at the roulette,"[58]

and concluding that not enough time had elapsed from the Big Bang to justify life appearance.

Marcelo Gleiser (2011), again based on a statistical subject, agrees that life might be distributed over the whole universe, but does not think the same for the intelligent life. He thinks that planets similar to Earth must be very rare, so postponing the problem to a possible future discovery of even a single extraterrestrial microbe. In his *A Tear at the Edge of Creation*, he writes,

"The proof of the existence of a cosmic imperative for life would take a hit to the Darwinian orthodoxy that discards all types of determinism in all processes related to life."[59]

Are we, therefore, compelled to accept that life is no other than a statistical accident? On the other side, if it is not a statistical accident and if it was not a relatively recent creative act, what was the *quid* able to let life arise and develop on Earth almost four billion years ago, taking into account the favorable conditions of the planet and the favorable numerical values of nature's constants? Does there exist a "sufficient condition for the explosion of life, able to support the final anthropic principle"?

2.6   What If Matter Were Engraved with Life?

From what is discussed, it seems that the only tangible argument able to find a condition "sufficient" for the emergence of life in the

---

[57]de Duve (1998, p. 31).
[58]Monod (1986, p. 141).
[59]Gleiser (2011, p. 396).

cosmos appears to have come from an event, unexpected from classical science, that occurred through an "emerging" property—the transmission of *information*[60]—along the self-organization path of matter. A long path whose beginning was the appearance of the most fundamental particles (electrons and quarks) and that continued with the construction of atoms and simple molecules (concretely derivable from the rigor of physical laws) until the formation and development of protein and DNA macromolecules and the following realization of many biochemical processes with mutual exchange of energy, matter, and information with the external environment and within the same cells. A property, the information, emerging at a very high level of complexity that realized the requested big jump of quality adequate to take charge of the rigorous and coordinated organization of myriads of cells, each one composed of about 10^{27} atoms, built like complex, work-producing machines, reciprocally supporting each other. A fundamental property, whose sudden appearance establishes the searched point of separation between (simple) inert matter and (complex) living matter and that must be considered intrinsic to matter itself, as it holds for the elementary laws of nature at a lower level of complexity.

As things are, more than thinking of a stroke of luck due to blind chance, *à la De Duve*, it seems that Gerald Schröder could be right when he writes,

> "[. . .] being in the need to face an enormous number of lucky stakes under the success of the evolution play, it would be a legitimate question that to ask ourselves until which point such success is inscribed in the structure of the universe."[61]

In the hypothesis that life may be inscribed in matter and considering that our solar system—in its origin and development—has physical characteristics similar to those of other billions of one-star planetary systems, we can dare to enlarge our look beyond

---

[60]As already discussed, the so-called "emerging" properties are connected to the nonlinearity of a system (e.g., Prigogine's quoted dissipative structures). A key role in these properties is played by the "collective variables"—resulting from the nonlinear interactions of the elementary components of the system (not always derivable from the knowledge of the elementary laws)—functional to the exchange of information and to coordinate the development of the system.

[61]Schröder (2002, p. 70).

Earth and beyond life as we know it. If life would simply derive from the tendency of matter to self-organize until the formation of very complex molecules, finally leading to the appearance of the *informative* property (like the formation of DNA), we will not need any statistical calculations to show the probability of occurrence of this event. If it is so, in fact, and taking into account the immensity of the cosmos with its billions of billions of stellar systems, made up of the same elements forged within the stars and distributed all over space, we may well think that the vital process may emerge everywhere in the universe, although perhaps in diverse configurations, such as to fit itself with the local physical-climatic conditions.[62] Like the human being on Earth, then, the process bringing life from inert matter could develop, in principle, on every other planet in any other conditions with specific adaptation paths—obviously with possible different results in terms of physical configuration and dimensions concerning living structures.

At this point a question arises: is life possible without carbon? Actually, even if our knowledge shows that the determinant element for life on Earth is carbon, it is also true that silicon and boron are as well able to form long and complex molecular chains. There could therefore exist, at least in principle, forms of life based on these elements, provided that they find favorable physical-climatic conditions, although different from those on Earth. John Barrow writes,

> "By modifying the twentieth decimal digit of the fine-structure constant's value, there will not be any negative consequences for life. But if we change the second decimal digit, the changes for life become much more significant. The atom's properties are altered and complex processes like protein's folding or DNA replication may be unfavorably affected."[63]

But, Barrow concludes,

> "who may exclude the possibility of new chemical combinations of which to-day is extremely difficult to evaluate future consequences?"[64]

---

[62]It would be a short cut on theological speculations of the traditional type. For a discussion of the relations between Christian theology and biology, see Boniolo (2003).
[63]Barrow (2004, p. 136).
[64]Ibid.

I agree with Barrow: why should we escape the problem? Is it necessary to imagine (in science as in religion) that all that we do not know is always and only in our image and similarity? Why do we not try to apply, here too, Occam's razor,[65] which has always been so useful for science? If it is true that the life phenomenon is due to a process of self-organization of always more complex structures, starting from the fundamental elements spouted from the Big Bang to the connected emersion of new properties not always derivable from the known laws, that life arises always, even in very unfavorable conditions (e.g., deserts or oceanic depths), that it may appear statistically impossible, for the enormous dimensions of the universe, that the small Earth is the only place containing such complex structures as living beings, is it not much simpler (in the sense of Occam's razor) to think that the capability of matter to produce life is inscribed in it?

If this would be the case, wouldn't life be to adapt to the environment (nature's constants, climate, etc.) and not vice versa? And, consequently, to have the possibility of different forms of life in worlds with different physical-climatic conditions? If this would be the case, the anthropic principle would lose its relevance and we would have only one condition, both necessary and sufficient, for the emersion and development of life, in whatever form it would exist—practically the same existence of matter with its implicit vital principle—and Jacques Monod would be right when he writes that

"[. . .] the accomplished structure is not pre-formed, as such, in any place, but its project is present in its constituents."[66]

Of course, we are making hypotheses; any possibility of verification is out of our present knowledge and technology, circumstances that allow us to explore only an extremely little part of our cosmic ambient.

Concerning life in our solar system, recent observations are starting to give news, particularly about elementary life in the

_____

[65]The "economy principle" in the interpretation of any phenomenon (Occam's razor) tells us that given a set of theories, explanations, hypotheses, or laws, the simplest is to be preferred under the same conditions.
[66]Monod (1986, p. 90).

underground of Mars and perhaps in the atmosphere of Encelado,[67] while we still have to wait—both in terms of observations from Earth and from new space missions—to have news about other planets and moons in our solar system. Who knows how long we must wait until we know what happens in the exoplanets recently discovered in other, close or far, stellar systems! Concerning, in particular, human beings, we don't know what the future reserves for us along our evolutionary path on Earth. We only know that man has developed an

"exaggerated sense of excel and dominate on Nature, so considering himself its unsurpassable vertex."[68]

James Lovelock, in his very recent book *Novacene. The Coming Age of Hyperintelligence*, thinks very unlikely the existence of other understanders of the cosmos other than us, in spite of the huge numbers of cosmic objects. As far as I understand, his certainties are almost referred to life as we know it on our earth and do not take into account the possibility of different aspects of life in function of different physical conditions encountered in other planets of other solar systems in our or in other galaxies. Neither does he take into account the possibility that life could be engraved with matter. In my opinion it is a big mistake to give an absolute value to the restricted knowledge we may have on our small earth.[69]

Perhaps we should ponder much more humbly on our smallness with respect to the cosmos and understand that nature is much richer and complex than we are, even considering our formidable intelligence. We should not ignore the possibility that—in the remote vastness of the universe or in the remaining four billion years of possible existence of our solar system, as it is now—nature's achievements significantly exceed both the fantasy and the intellectual abilities of a human being. De Duve writes,

"there are no reasons for which we have to consider ourselves the top of a process that has still five billions years available [. . .]."[70]

---

[67]I refer here to the discovery of organic molecules within the geysers of Encelado, Saturn's satellite.

[68]de Duve (1998, p. 15).

[69]Lovelock (2019).

[70]Ibid.

Chapter 3

The Evolution of the Cognitive Experience

The way we see the world is affected by the experience we have of it, constantly elaborated by our brain. At the dawn of civilization, the acquisition and elaboration of our experiential data were almost all based on an unconscious process of resonance between man and environment, a process amplified by our body and transmitted to the emerging conscience. A knowledge modality, this (made of empathy,[71] envy, passion, fear, contempt, attraction, hope, etc.), with both positive and negative features for social cohesion, and where the flow of time was anticipated by a continuous present. A modality, this, that today is identified as "irrational," a term often used—not always correctly—in a negative sense.[72] We may have a good idea about the emphatic-participatory epoch, here quoted, by studying anthropologists' reports on human groups located in regions not interacting with our modern society, and then basically differing from us in their absence of writing. Levi-Strauss writes,

---

[71]About the concept of "empathy," see Boella (2018). The topic is discussed from the neurologic viewpoint by Rizzolatti and Sinigaglia (2006).
[72]This discussion is a re-elaboration from Alfano and Buccheri (2006, pp. 269–302).

Myth, Chaos, and Certainty: Notes on Cosmos, Life, and Knowledge
Rosolino Buccheri
Copyright © 2021 Jenny Stanford Publishing Pte. Ltd.
ISBN 978-981-4877-33-6 (Hardcover), 978-1-003-08869-1 (eBook)
www.jennystanford.com

"These societies offer the only model to understand how humans have lived in an epoch corresponding, no doubt, to the 99% of the total duration of the humanity's life and, from the geographic viewpoint, on the three quarters of the Earth's inhabited surface."[73]

During evolution, the need to communicate gave rise to the oral language, initially of mimetic-poetic type, based on mythical images, those that Socrates considered as coming from divine inspiration (poetic "mania"):

"Who comes at the doors of poetry thinking that he will certainly become a good poet, remains incomplete, and the poetry of those remaining in mind will be obscured by those taken by mania."[74]

Subsequently, orality acquired the characteristics of a dialectic type, whose question-to-answer structure started to show a spatial and temporal separation between the communicating people.[75] With the writing, appearing after orality, every residual space-temporal superposition between the emitter and the receiver of the communication was eliminated, with the result of making always more difficult the emphatic-participatory resonances and of going toward a more quantitative and objective interaction.

With the incoming orality, a rational way of thinking started to develop. Stabilized by the writing, the rational thought is today mainly based on the categories of logic, mathematics, and empiricism, typical of the scientific approach to knowledge, and on a more precise consciousness of the flow of time from past to future. The "subjective" participatory attitude, anyway (accrued during the previous centuries and alimented by the unconscious knowledge), continued to be present and pressing, although in the background, and was prepared to be integrated with the logics instruments. An essential integration, this, both at the emotional and at the practical level, for the full satisfaction of our own tendencies and needs and for the solution of problems otherwise unresolvable.

---

[73]Levi-Strauss (2017, p. 28).
[74]Platone (1998, p. 63).
[75]Ong (1986, pp. 23–28).

3.1   Myth and Archetype

The precise moment in which man became able to transform to symbols his experiences, later communicated through mythic tales, is not known. According to Erich Neumann, however, it is just the myth—in its central themes found in all cultures of the ancient world—that reveals the first steps of the process of the phylogenetic development of the self. It seems certain that the presence of high-level feelings, like solidarity and "moral" sense, was already established in primates, as we know from the ethologist and primatologist Frans de Waal. By reading his book *The Bonobo and the Atheist*, in fact, we find that animals manifest empathy and moral tendencies as human beings do,[76] suggesting that morality is not a human invention as we like to think.

Coming back to humans, it is known that the degree of complexity of the Cro-Magnon's brain was already so high as to consent to the presence of a forcefully active mind by means of its properties, for example, the consciousness of existence and the ability to connect his own fears and joys to symbols rich with emotion, at the cost of referring any experienced but uncontrollable events to beings endowed with supernatural power. Myth was already born. Myth that Aristoteles considered as the base of man's desire to know the deep motivations of what he observed in nature:

> "who feels a sense of doubt and wonder acknowledges to not know; for this, even who loves myth is somehow philosopher: myth, actually, is made up of things arousing wonder: so, if men have philosophized to get free of ignorance, evidently they try to know just for knowledge."[77]

After a long period in which the myth had been relegated to a role of secondary importance because it was considered incompatible with the laws ruling the universe, Karl Gustav Jung, by referring to the most ancient cultural and religious fonts of man's history, led us to full comprehension of the myth's psychologic role by elaborating on the theory of archetypes. Károl Kerényi writes,

---

[76]de Waal (2017, pp. 35–38).
[77]Aristoteles (A 2, 982b, 12–20).

"Myth, in a primitive society, that is in its original live form, is not simply the report of a story, but a lived reality [. . .] a manifestation of a primary, superior and of higher importance reality that determines life, destiny and humanity's present activities, while men extract from it both motivations for present rituals and moral acts, and warnings on how to put them in practice."[78]

According to Jung, the archetypes of the collective unconscious are the psyche's structures common to all mankind that enter unconsciously in the lives of everybody, unlike the repressed personal or frustration memories, like childhood fears and traumas, that Jung defines as "personal unconscious." So, as archetypes are common to the human species, the personal unconscious is biographic; it holds for the social life of single individuals. Joseph Campbell writes that man

"[. . .] has an inherited and a personal biology [. . .] The archetypes of the unconscious are expression of the first [. . .] The psyche's deepest layer expresses the instinctive system of our species that has its roots in the human body, in the nervous system and in our wonderful brain."[79]

The deep truths expressed by myth that filled the primitives' minds with information stored fundamentally with a "symmetric logics,"[80] so giving rise to archetypes common to all cultures of the world, are seen by Kerényi as models of symbolic thought. He writes,

"Modelling in mythology is imaginative. It is a 'springing' and, at the same time, an 'unfolding': it is fixed as all mythologemes are fixed in the sacred traditions, it is a sort of artwork. There can be various different developments of the same fundamental theme, one close to the other or one after the other, like the diverse variations of a musical theme [. . .] But always with deeds: that means with something objective, something already become autonomous, speaking for itself, something to whom is not done justice with interpretations and explanations, but only keeping it in mind and leaving that it expresses its own sense [. . .] Mythology is not only a way of expression at which place we could choose another one, simpler and more comprehensible."[81]

---

[78]Jung and Kerényi (2012, p. 19).

[79]Campbell (2007, p. 84).

[80]The symmetric logic, defined by Ignacio Matte Blanco, will be briefly discussed in subsequent chapters.

[81]Jung and Kerényi (2012, p. 16).

3.2 The "Nuclear Conscience"

At the origins of the evolution of the human thought, a knowledge modality of effusive-participatory type was active, typical of a close man–environment interaction and manifested through a poetic-symbolic language. A kind of knowledge, this, attenuated today due to the emersion of a rational consciousness where the appreciation of qualitative aspects has gradually reduced in favor of quantitative ones, with a simultaneous greater separation of the observer from the environment.[82]

Such a primitive knowledge modality, appearing at the dawn of the onto- and phylogenetic man's evolution, is indicated by Antonio Damasio as a "nuclear conscience" that works

> "when the cerebral devices generate a non-verbal description, through images, of the way in which the status of the organism is modified by its elaboration of an object [. . .]."[83]

On the anatomic level, the functions of the nuclear conscience are referred to the right cerebral hemisphere, the center of gestural language and acting place of the visual-emotional-imagination, of creativity, of intuition, and of the parallel elaboration of information, a hemisphere that has an important role in the first phases of life, that is, in man's infancy (onto- and phylogenetic), when, as Elkhonon Goldberg writes, "the arsenal of ready-to-use models is still limited."[84]

With his formulation, Damasio tries to unify mind, brain, and body on the basis of rigorously scientific data, by stigmatizing Descartes' error, consisting of the idea of a pure thought, of a rationality not affected by emotions and feelings.

According to Damasio, Descartes did not know that the apparatus of rationality is not independent from that of biologic regulation, and that emotions and feelings are often able to strongly affect, unbeknownst to us, our beliefs and our choices. Our mind is not structured to present to us a list of rational items, favorable or against a given choice. It, instead, acts in a much faster way (although less precise), taking into consideration the emotional weight of our

---

[82]Alfano and Buccheri (2009, p. 97).
[83]Damasio (2003, pp. 105–133, 237).
[84]Goldberg (2005, p. 193).

previous experiences and giving out an answer in form of visceral sensation. Damasio writes,

> "There is a diffuse belief that the use of logic may be intrinsically able to lead us to the best solution among the ones possible for any problem. An important aspect of such a rationalistic belief is that, in order to obtain the best results we should exclude emotions [. . .]"

and again,

> "Our brain may often decide well in minutes of fractions of minutes, according to the temporal reference from us established as appropriate for the objectives we have in mind; if it is so, it must execute its admirable task not only with the pure raison. Another perspective is needed."[85]

The nuclear conscience, therefore, bound according to Damasio to primordial instincts, perceives and stores in the unconscious the experiential data in a global and indistinct way. Data, these, that are lately recoverable all of sudden, "intuitively," when, under the urgent stimulation of the environment, it is necessary to solve, without any rational analysis, problems not solvable in other ways. Daring a parallel with the quantum physics, Jauch writes,

> "Just as the correlations between interacting quantum systems introduce in the behavior of single systems many different types of changes (that, lacking a better term, we call 'quantum jumps'), in the same way any person integrated in a group shows spontaneous intuitions that would be inaccessible in the isolation,"[86]

It has to be stressed that the kind of knowledge corresponding to the nuclear conscience cannot be objectified, since it is subject to the personal interpretation of the same percipient, and often it is not even possible to express it with words in an absolutely clear way. For these reasons, this knowledge modality is characterized as "irrational." We can ascribe to it, in particular, the intuitions, sometimes ingenious, of artists and scientists, as well as spiritual expressions of mystics and saints recalling the revelation of the divinity.[87]

---

[85]Damasio (2008, pp. 242–244).
[86]Jauch (2001, p. 124).
[87]Hillman (2005, p. 13).

3.3 From *Mythos* to *Logos*: The "Extended Conscience"

The birth of a rational conscience and the use of a linear, apophantic language, typical of today's scientific attitude, have established a perspective's shift, described by neurophysiology, with the displacement of the mental control from the right hemisphere to the left one of an already experienced phenomenon. Goldberg, on the basis of imaging experiments on the activation of various regions of the brain under stimulation, supports that the left hemisphere is responsible for most processes based on the models' recognition, involving or not the language. He writes,

> "An individual who faces a truly new situation or problem, appeals mainly to his right hemisphere. But, once the situation becomes familiar and known, the dominant role of the left hemisphere becomes evident." It looks that all models in which the essence of the situations is epitomized [...] are stored, once formed, in the left hemisphere."[88]

Such a displacement is due to the emergence, in the course of time, of a system of thought, juxtaposing to the nuclear conscience already existing, and defined by Damasio as "extended conscience," a system for storing and re-elaboration of single, instantaneous units of nuclear conscience that has increased both in the course of human evolution and during the life of any single individual, until including not only the "here and now" but also the prevision of the future and

> "all that part of the past needed to effectively light up the hours [...]. The extended conscience include totally the 'nuclear conscience' but is better and greater and grows in the course of time, experience after experience."[89]

We may refer to the extended conscience any knowledge of logic-scientific type for its capability to save and catalogue the myriads of experiences done thanks to nuclear conscience, and possibly to retrieve and exhibit them by means of the linear language. The rational activity supported by the superior cortical structures is fully

---

[88]Goldberg (2005, p. 186).
[89]Damasio (2003, pp. 237–238).

conflictual with respect to the primeval form of nuclear conscience, the difference being the fact that the subject has assumed, in this case, an external intellectual position, not interacting with the object of its observation. From the anatomic point of view, the extended conscience is represented in the posterior paralimbic and cortical regions, depending very much on the cerebral cortex, particularly the associative cortex, and relies mainly on the functional contribution of language zones—Broca's and Wernicke's areas—of the left cerebral hemisphere, especially the prefrontal lobes.[90]

With the advent of rationality, the myth and its symbolic expressions, inexplicably intertwined with bad and good, with wrong and correct a priori beliefs, did not disappear and, perhaps, may never disappear. Actually, even opposing myth by considering it incompatible with the apophantic speech—which is structured according to the rules of logics—the reason, after having become aware of the evolutionary stories from where it derives, re-evaluated the myth and acknowledged its importance as a primary form of knowledge, a not-eliminable intellectual man's attitude concerning the external reality. A myth that, although irreducible to reason, is able to integrate it, sometimes enriching it, other times implying internal contradictions. An attitude, the mythical one, implying for the reason the burden to manage the integration and the awareness that sometimes it will be thwarted and sometimes accelerated in a succession of proofs and errors, follow-ups and regressions, that the formed man's duality produces in the unceasing encounter-collision between the two knowledge modalities. A myth, therefore, that reason, subsequently intervened, tries to explain and analyze with the aim of keeping under control its variability and to be able to use it as an additional resource toward the progress of knowledge. Campbell writes,

> "Ancient taboos based on old mythologies have been shattered by modern science [. . .] the conscientious teacher worrying about the moral and cultural education of his students, must be loyal to myths supporting our civilization or to truths verified by science? Or, perhaps, does exist a wisdom element beyond the conflict between illusion and reality, thanks to which our life could still be recomposed?"[91]

---

[90]Idem, pp. 237, 282.
[91]Campbell (2007, pp. 12–13).

It is true that mythical tales are false if we take them for truly happened events, but if, as Jung does, these tales are analyzed like mind's expressions transformed in symbols, they

"help mankind to appreciate both the external reality world and the interior reality of each of us [. . .] we do not need to be anchored to those emotional schemes and archaic thoughts, inappropriate to our contemporary life."[92]

In any case, we need to establish

"a dialogue by means of symbolic forms elaborated by our unconscious and recognized by conscience in a continuous interaction."[93]

A necessary interactive dialogue, therefore, is needed between the fortuity with which the symbolic forms come out from the unconscious, as dictated by our own needs, and reason's rigor keeping them under control, as it happens in the evolution of cosmos and of life, where the rigor of the physical-chemical and adaptation laws, respectively, keep under control all apparently random configurations, looking far to the objective of a coherent development.

3.4   Two Cerebral Hemispheres, Two Logics

The asymmetric brain's organization already discussed, with two hemispheres reciprocally interacting through a vast and complex network of nervous paths, timed within the *corpus callosus*, shows a duality of opposite and reciprocally irreducible modalities for the elaboration of information needed by us for observing and analyzing our environment.

The possibility that such a dual characteristic of the brain could correspond to the theoretical system proposed by Ignacio Matte Blanco has been discussed in the literature: a system in which two opposite and reciprocally competing "logics" collaborate with each other, aiming to rule the storage and analysis of the information exchanged with the environment, in view of our best-

---

[92]Idem, p. 14.
[93]Idem, p. 17.

possible adaptation to the external ambient. A sort of *bilogic* made up of an "asymmetric" logic, mostly referring to recently acquired data, stored and analyzed by our rational system of thought, and a "symmetric" logic, mostly referring to older information, hidden in the deep unconscious and stored according to the principles of "generalization" and "symmetry," defined as follows. For the generalization principle,

> "the unconscious system deals with each individual object (or person, or concept) as if it were a member or elements of a large class containing other members, and deals with this class as it were a subclass of a more general class etc. [...]."[94]

This principle is frequently applied together with another principle for which

> "among all the possibilities of generalization offered, only some are chosen and some are left out. So, some individual characteristics from which the generalization started remain frequently in the general class as it is finally formulated."[95]

The symmetry principle says that

> "the unconscious system deals with the inverse of a relation as if were identical to it. In other words, it deals with asymmetric relations as if they were symmetric."[96]

From this principle we deduce the important consequence of the nullification of the concept of order.[97] A circumstance, this, implying the absence of the temporal process. Furthermore, the generalization principle, together with the symmetry principle, produces the exchange of an object/person with another of the same class or subclass, for example, the exchange of a mother (class of persons who nourish materially) with a teacher (class of persons who nourish mentally), both belonging to the more general class of persons nourishing other persons. Lucia Figà-Talamanca Dore writes,

---

[94]Matte Blanco (2000, pp. 43–44).
[95]Ibid.
[96]Ibid.
[97]Figà-Talamanca Dore (1978, p. 19).

"By exchanging a mother with a professor, the unconscious does not limit itself to deal these two concepts as having something in common according to the generalization principle, [. . .] does something more, deals with them as they were identical, because, for the symmetry principle 'if x is part of y, than y is part of x'."[98]

The unexpected information coming almost spontaneously from the unconscious, in obedience to these principles, cannot be generally elaborated with the usual instruments of the linear logics. A thesis, this, certified both from the configuration of some dreams apparently incomprehensible according to the asymmetric logic of the waking state and from the often very difficult translatability in words of our deep thoughts following the rational schemes of the apophantic discourse, at variance with what happens for any information recently acquired, whose logic is maintained when returning to the conscience state.[99]

In this detection of bilogic structures, Matte Blanco solves the Freudian difference between conscious and unconscious, bringing the last to adhere to the structures of myth and then contributing to give the correct placing

"to all events of emotional type, fonts of inexhaustible enriching of the human experience, in their role of creative potential's expansion operating in various aspects of the thought."[100]

For its encompassing way to conceive the *logos*, the bilogic structure allows us to recognize that the final judgment role has to be given to the rational component of the thought—called "asymmetric thought" by Blanco—that Pietro Bria and Fiorangela Oneroso, in their introduction to *Bi-logic between Myth and Literature*, consider

"able to seize the deep foundation of knowledge and of reality, and to recognize the difference between unity and multiplicity, simultaneity and succession, co-presence and displacement, so leaving to asymmetry the role of a pure out-of-conscience event [. . .] with respect to which the 'symmetric thought' is anyway debtor for its way to be expressible in words, that is in order to have the possibility to secede from that

---

[98]Idem, pp. 20–21.
[99]Alfano and Buccheri (2012, pp. 256–257).
[100]Bria and Oneroso (2015, p. 13).

whole in which is merged in origin, and to get free from the situation of constituting disorder, thanks to the forms always more developed of the logic-rational thought, able to explain and translate it."[101]

3.5 A Mixing of Unpredictable Results in the "Project" of the Living

Already at the beginning of its formation, unavoidably bound to the terrestrial environment during life's evolution, the "project"' of each single individual—written in his/her DNA—even similar to the others at a programmatic level, is revealed to be an *unicum*, never identical to that of any other individual, thus implying that not all the sensory organs of a person have the same identical perceptive and descriptive ability against the external reality. It is, vice versa, reasonable to think that the (even small) major or minor sensitivity of a sensory organ with respect to another one could imply the attribution of a different value to stimulations coming from that organ[102] and therefore a position of greater of smaller importance to the data admitted to analysis by our brain.

The consequence is that every small difference in the selection dynamics of external stimuli may produce, in every person (and at a higher level in every human community), corresponding differences of weighing the interpretation of the single cognitive units and of their reciprocal relationships representing the external reality. To this singularly defined response of our sensory organs, derived from our own specific "project" connected to our specific physical and psychical structure, we have to add all other factors that feed the variability of our experience of the world, independently of the fact of living in the same social environment. Examples of this variability are our personal relationship with the physical or economic conditions, our precomprehensions, and all those personal prejudices raised up along our life and slowly transformed in hard schemas of thought

---

[101]Idem, pp. 13–14.

[102]It looks reasonable, for example, that the difference between the pleasantness and painfulness of perceptions are not equally felt by all of us. It is known, in particular, that some painful perceptions may, in some cases, be eliminated from the conscious memory, with the possible consequent emersion of alterations of the behavior, due to an unconscious influence of their presence.

to which we refer to any time we observe or interpret facts and phenomena.

It is always the "project" in its unicity that oversees, with its specific way of mixing of the two cognitive modalities (the instinctual and the rational ones), both the integration of the data extracted from the unconscious with those coming from the present experience and their subsequent analysis. An integrated mixing of intuitive perceptions and experiential data that allows everyone to face the variety of life conditions in a unique and unrepeatable way, so determining the unpredictability of the result. The *unicity* of our self-consciousness derives just from these differences of perception, evaluation, and successive behavior between individuals. An asset that, being strongly bound to the hosting body, shapes its development in a precise direction, different from that of other individuals but anyway leading to satisfy his/her own vital needs.

The scientist-theologian Michał Heller proposes three aspects of the mixing of rationality with increasing presence of the irrational modality[103]: (i) the rational structure of the universe, accessible with the mathematical-empirical investigation of nature adopted by science; (ii) the reasoning, that is, the use in our conversations of "logic" instruments acting to derive correct conclusions from clearly defined premises; and (iii) the "rational" way of living, where the rules, although in accordance with those of the reasoning, are very difficult to encode because of the many deviating factors discussed before, including our life's objectives and the means used to reach them.

It is easy to understand that the three aspects are connected to each other. Heller notices, in particular, that the perfection level of the rational system of thought decreases gradually—and correspondingly increases the presence of the "irrational" aspect— when going from the first to the third aspect. It is, in fact, true that our way of living is able to affect somehow our reasoning as well as the way we investigate nature. In this last case, however, the influence due to irrational aspects is very low because of the powerful collective approach of science, where the critical verifications from

---

[103]Heller (2012, pp. 287–290). In his book, Heller talks about the "level of perfection" of the rational system of thought, while here, coherently with what we are discussing, I have preferred the equivalent concept of "level of mixing" of the two interacting knowledge modalities.

the scientific community, obtained through repeated observations and experiments, contribute to almost nullify any individual difference in beliefs or cultural traditions among scientists and, vice versa, to strengthen any positive feeling. Sometimes, just one good intuition by a single researcher, carefully and repeatedly verified by the community, is enough to guide the knowledge process toward the correct direction. The great achievements of the scientific method, able to build very precise models of the directly observable reality, witness its general effectiveness.

The mediating interaction by many persons, able to activate shared verifications, is possible also in the second and in the third aspect considered by Heller, but due to the large number of unaware and unverifiable preconceptions, its effect—certainly greater in the case of reasoning than in the everyday behavior—is anyway much lower than in the investigation of nature. In particular, concerning the second aspect discussed by Heller—reasoning—the effectiveness of preconceptions generally decreases in the course of our life experience because of the mediating interaction of many different persons. Unfortunately, this is not always true; in fact, when an idea—incorrect but able, in particular social conditions, to attract the attention of large masses of persons—is proposed and supported by powerful influencers or by powerful interested groups, the interaction process may take very negative directions, thus leading to disastrous situations lasting long over. Situations like this have happened very often in human history, and continue to happen today, as shown, in particular, in the present social and political situation in Italy as well as in many other countries of the world.

3.6 The Mental Models of Reality: Filters, Ambiguities, Contradictions

The continuous elaboration of the information derived by our life experiences, in all their aspects, results in building up a suitable organized representation in our mind. It happens that the imprint of objects or events with their physical, psychologic, and practical properties—size, dimensions, color, use, feeling, etc.—is stored in our brain, starting from the stimuli transmitted from the sensory organs (eyes, ears, nose, skin, taste buds) and from the following analysis done, also in relation to similar stimuli in the past. All these "cognitive units" about objects and events related to our daily experience are put together in an integrated mental construction, an individual global view of the overall situation, of which we are aware only partially, that has been called the mental model of reality (MMR)[104]. An interpretative scenario, the MMR, bound to the experiential specificities of each single person in consideration of the unicity of the "project" cited before, that continuously updates itself with always new observations and new mental elaborations. In building the MMR, both the cognitive modalities intervene with their own individual peculiarities, respectively, guided by our two cerebral hemispheres: the older one,[105] with the data acquired by intuition, tradition, or emotional, political, or religious belief, following a "symmetric" logic process,[106] and that of rational nature, more recently formed, where every proposition follows from the previous one according to a rigorous logic process, the one that Blanco defined as "asymmetric" logic.

In the MMR, even if reciprocally untranslatable, the two modalities mix together in a variable ratio from person to person and in the course of time, taking into account any new data continuously acquired, thus resulting in a corresponding gradual adjustment of our social behavior. A situation, this, not easily predictable because of the activity of our unconscious that influences our wishes, fears, needs, precomprehensions, beliefs, and prejudices of any kind that continuously infiltrate the MMR, so affecting our reasoning and its connected behavior.

The MMR is not a static structure; it is affected during life by continuous, temporary instabilities, following the interaction with other individuals and with the environment. Instabilities that imply unceasing modifications that affect, in turn, the MMRs of other individuals. This interactive process is neither linear nor equal for everybody and shows important analogies with the "chaotic"

---

[104]I refer here to Kenneth Craik's studies (1943) and to their subsequent elaborations made by Philip Johnson-Laird, who worked out a theory of those cognitive processes indicated with the term "mental models."

[105]As a result of the previous discussion, the terms "mythic" and "irrational" will be used interchangeably in some cases.

[106]Those elaborated, according to Matte Blanco, following a "symmetric" logic.

processes because of the occasional presence of bifurcation points, implying the need of sudden choices to which we respond according to our contingent state (physical, mental, economic, . . .). Choices that will certainly entail future ordering potentials but also new possible ambiguities, both internal and in the reciprocal interaction. For this reason and due to the influence of our own individual "project" in the analysis of sensory data, the MMR of each single person can never be identical to that of any other person.

Another important feature of all MMRs is the need to internally stabilize inconsistencies and contradictions in order to minimize any risks for the related psycho-physical structure, always possible in view of the often contradictory life experiences. In a way, we somehow force ourselves to interpret the external events—that can be rich with true and false knowledge—in accordance with our psychical conditions. A circumstance, this, implying both positive and negative connotations for each individual and for the whole society, in dependence of what we get from our unconscious and from the outside, and how we integrate them with the continuously updating information.

Among the positive connotations of this mixing, we may quote those important conquests of thought obtained by integrating dazzling intuitions within the rigorous paths of reasoning, including the many intuitions that help us in any aspects and moments of life, to solve pressing problems, not easily feasible with only the conscious rational thought.

Our "fishing" into our unconscious, however, is not always positive for ourselves or for society. Generally, when looking for information useful to the solution of our problems, the answer we get is more easily and relentlessly in strict relation to our psycho-physical structure and possibly in contrast with our social interaction. An answer, therefore, that contains the risk to stimulate conflicts between our own "cognitive units" and then halve dangerous and oppressing contradictions within the overall MMR structure, especially considering that the new acquired knowledge does not have always a satisfying level of compatibility with that already present in it. Fortunately, the usual incompleteness of the communication processes—both in quantity and in quality—helps us, since for sake of brevity, the formal language is normally accompanied by nonformal codes (gestures and other nonverbal communication, not speaking of the large variety of nuances of meaning, proper of all languages), which allows a not univocal interpretation, able to drastically reduce the dangerousness of many contradictions. Real filters to the communication, the last, that the audio-psycho-phonologist Alfred Tomatis discovered and studied in their neurophysiologic aspects by measuring the "listening curves" that characterize the audio-vocal circuits of many communities, all invariably and surprisingly different from each other.[107]

Most of the time, these filters allow us to adopt a favorable interpretation of any communication, such as to let it result compatible with our own present structure and so eliminate the danger. This is a soothing result for our personal psychology, since it allows us to filter—often unknowingly—any new information, possibly misrepresenting it, but anyway able to adapt it to our points of view and to our beliefs, even at the risk to choose only what is comfortable, so eluding any contradictions that could become very dangerous for our psycho-physical integrity and insomuch solvable only by a partial or total removal of the conscious memory. It is a sort of confirmation that the external reality remains for us unknown in its fine details and that it is almost impossible to have a completely sharable description of the world in all its innumerable nuances, although we continue to think it independent from us.

This, anyway, does not equal to say—as the idealists like Berkeley or Leibniz did—that the reality is only a construction of our mind, but whatever its ultimate essence would be, reality exists and is perceivable by everybody, but anyway with different and univocally imagined details by each one of us.

Martin Heidegger was conscious about these cognitive limitations and wrote that an "objective" knowledge is impossible since every knowledge is the result of an "interpretation" on the basis of our predispositions, previsions, and precognitions. According to Heidegger, actually, the act of interpretation is never the result of a neutral knowledge of something "objective" but only a representation of an interpreting subject. The reality is never precisely measurable and may be revealed only partially by humans.[108] Such a concept was shared by Hans George Gadamer, for whom the process of increasing knowledge consists of a "hermeneutic circle" where we have already

---

[107]Tomatis (1993).
[108]Heidegger (2000, pp. 642–670).

some knowledge of what we are going to learn. Our comprehension, then, cannot be reduced to a simply reproductive act; it has instead a poietic, productive component.[109]

Being aware and accepting such a view without amplifying its individual value is a way to encourage the dialogue between persons, particularly between religious faiths, among the most difficult and controversial, individual and social, confrontations on the planet.

---

[109]Gadamer (1983, pp. 342–357).

Part II
Within Society between
Mythos and *Logos*

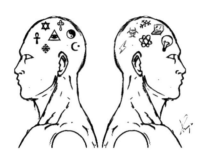

Within Society

Active, brilliant, inquisitive Tom,
in quality and quantity makes and shares,
leaving to others surpluses undue.
Gloomy, lazy, nosey Dick,
out of envy, Tom watches and tracks;
for others' openness politely holds.

Too much ease, culture kills,
below the minimum, life blows out;
"in media stat virtus" for a fair society.

—Rosolino Buccheri, 2011

Premise: Antinomies and their contribution to knowledge

The term "antinomy" recalls the presence of two conflicting concepts within the same set of accepted rules, their meaning appearing as much abstruse as much precisely we define the set of rules that we refer to. In particular, if we stay within the rules of symbolic logic—the discipline defining the procedures to follow in logic reasoning—an antinomy will appear as a real logic contradiction, and therefore all incomprehensible, and we will be in presence of an antinomy *stricto sensu*. If, more generally, we refer to the set of rules and conventions (written or not) referring to traditions, moral rules, esthetic sense, and whatever else characterizes our social behavior, we can more properly talk of contradictions or, in more serious cases, of paradoxes (from Greek παρά and δόξα), that is, descriptions of facts that contradict our common opinion or our daily experience, or argumentations leading to not acceptable inconsistencies.

In Immanuel Kant's philosophy, antinomies are the consequence of our ignorance about the basic premises to a proposition; to solve them, it is enough to include the premises, if known, or to make higher the hierarchic level of those propositions that seem to be antithetical at a lower level.[1]

Zeno's paradoxes, described in the fifth century and quoted by Aristoteles in *Physics*, are antinomies *strictu senso*. The first, the "stadium paradox," says that a runner takes an infinite time to reach the end of the stadium because he/she has firstly to reach its half, but before reaching the first half, he/she has to reach the half of the half, and so on, which appears to mean that the end will never be reached. The second, "Achilles and the turtle," affirms that Achilles' "fast foot" would never reach a turtle that started earlier and with an advantage over him, because when he would reach the position originally taken by the turtle, it, in the meantime, would have reached a new position, and so on; their distance, therefore, however becoming always smaller would never cancel. Zeno's paradoxes were overcome only in the seventeenth century, when the infinitesimals

---

[1]The first of Kant's antinomy consisted of the thesis that the world was limited in space and had a temporal beginning, in contrast to the antithesis for which the world was infinite both in space and in time. Such an antinomy, as well as others of the same kind, was due, according to Kant, to our belief that both thesis and antithesis could be good representations of reality. If verification could be possible, any contrast would disappear and only one of the two arguments would be true, the other becoming false.

were introduced in mathematics, together with the observation that the sum of infinitesimals may result in a finite number.

Incidentally, a very interesting paradoxical situation is described in the literature by the Reverend Abbott in his tale *Flatland*, when a bidimensional storyteller becomes aware of the misunderstandings arising when visiting a unidimensional world and, successively, by being caught in paradoxical terror when realizing the appearance in his world of a tridimensional visitor.[2]

As Kant observes, every paradox may possibly be solved by imagining to look at the event under study from a higher dimension site, as is the case of the geometrical antinomy on God's presence in the world, as described in Dante's *Paradise*, which is solved within the modern concept of the mathematical hypersphere, a sphere with more than three dimensions.[3]

The assumption of the existence of a hierarchy of "logics" is not so farfetched, and the presence in us of two knowledge modalities, where the "symmetric" one shows to be somehow comprehensive of the other, the "asymmetric" modality, is a clear indication of it. Actually, our individual mental models were primarily formed by the raw backlog of all the evaluations derived from the daily experiences made, based on the physical co-oscillation of our own body with the environment during the mythic phase of humanity, as discussed before. In those conditions, where a "linear" logic was not yet present, neither antinomies nor logic contradictions could be perceived, and our forefathers reacted to them (as already acutely recognized by Lévy-Bruhl)[4] by means of an "affective category of the supernatural" to the complexity of any informative content coming from the outside, making them acceptable by primitives "with our same live and net awareness and belief."

With the increase in information and with the evolution of superior cortical forms, experiences stored started to be connected to one another, showing the rise of a "linear" logic, favoring the perception of antinomies, as soon as new phenomena or events appeared, not compatible with the scenario already existing in our mental models.

_____

[2]Abbott (1998).

[3]A very interesting comparison between Dante's poetic imagination and the mathematical concept of hypersphere is discussed by Osserman (1997, pp. 93–95).

[4]Lévy-Bruhl (1973, pp. 19–20).

On this basis, it is presumable that the reason may have always had the possibility to harmonize the contents of the two modalities and that it is always possible to reduce to *logos* many contradictions inexplicable at first sight, provided one has the courage and humility to rise to a superior level. Courage that helps avoiding to be stiffened by the presumptuous inertia of considering fixed and unchangeable our present knowledge and, instead, subduing to the humble application of the neutrality principle when listening to others, humility that makes us flexible with respect to the expansion possibility of our mental models.

We could include among the antinomies *latu senso* all those situations appearing to us paradoxical and then irreconcilable with the internal logic of our mental models of reality (MMRs). They may be caused by the simultaneous presence of our two opposite and mutually irreducible knowledge modalities, which lets them appear as paradoxes of a dynamic type, that is, due to variable life's experiences. Also in these cases, in agreement with Kant, the solution may be found at a higher hierarchic level, where our cerebral duality, although independent from any communication rules, is subject to the control of the linear dynamics of our reason, of which mathematics and formal logic are emblematic representations.[5]

Despite the great informative material provided by science in more than 2000 years of research—knowledge for which we have been able to recognize and overcome so many paradoxes—the basic antinomies, those relevant to the abysses of our unconscious, shall still have to wait, at least until the neurosciences investigations will definitely shed light enough to decode the way our cerebral activities work, thus relevantly contributing to advancements of psychological theories and allowing us to look at the antinomies within a suitable superior descriptive logic.

Opportunity, this, not taken into consideration for a long time by science, both because of being affected for centuries by a monoperspective linear strategy and because of looking at the

---

[5]In support of Damasio's theses, I believe, together with many others, that the kingdom of mathematics lies not in the skies, as Penrose claims (Penrose, 1999), but that mathematics is just a great product of human intelligence (not necessarily an ultimate one) on which, in the course of the past centuries, the social consensus plateaued.

paradox from an amusing point of view, thus lacking the occasion to understand its origin and the possibility of a solution. Such attitude, unable to explain all natural phenomena through their effective reduction within elementary schemas, is still now an obstacle for many who do not consider interesting and worthy of deep attention everything that remains out of a schema describable by a linear logic.

Fortunately, it is commonly accepted that the most important task for science—if it does not want to limit itself to the obvious technical deepening—is just the task to explain all what does not appear "logic" at first sight. An explanation that could even occur at the expenses of radically modifying our own strongholds, which, being established by man, may not necessarily be absolute truths but simple work hypotheses from which to start in view of going always deeper in nature's comprehension so as to use its resources in an economical and respectful way for both, nature and man.[6]

A neutrality principle, the last, by which to get inspired in order to never take for final even the greatest thought's achievements. Science is a product of man, who is a product of nature, and therefore it is our science that has to conform to nature, not vice versa. Even the most exact of the experimental sciences, physics, our product, had to succumb to the evidence of the presence of the antinomies. At the beginning of the twentieth century, when everything looked to be subject to the hard, precise, and perfectly comprehensible laws of the realist determinism and somebody just dreamed to have already "decoded God's mind"[7] by means of a thought "theory of everything," the physical investigation of the microcosm, by means of quantum theory and by studies on complex systems, declared the "end of certitudes,"[8] so leading us to a no-return way, with the evidence that nature cannot be described in the rigid terms imagined by the classic realism.

---

[6]Friedrich Dürrenmatt writes "[. . .] work hypotheses have to conform to the human being; human being have to conform to clear truths" (Dürrenmatt, 2003, p. 9).
[7]Stephen Hawking wrote: "[. . .] if we shall get to discover a complete theory [. . .] we will decree the final triumph of the reason because we shall know God's mind" (Hawking, 1997, p. 197).
[8]Prigogine (1997).

Actually, photons, electrons, protons, and all the large zoos of "elementary" particles,[9] studied by the physics of the "infinitely small," have a dual nature. According to the way we observe them and to the instruments we use, they show themselves to us with antithetic characteristics, all incompatible with the logic of the realism, for which (as an example) an object cannot be precisely localized in space and, at the same time, be infinitely extended like a wave. Despite such counterintuitive descriptions, the observed duality cannot be disproved—and therefore quantum theories cannot be wrong—if we consider the enormous quantity of phenomena explained with a mathematical precision never reached before.

---

[9]The term "elementary" given to the gradually discovered particles has shown to be a too simplistic one, even if very effective at a practical level; truly elementary particles, in the sense given by Greek atomists, do not exist, and the wave–particle duality is a probative confirmation of it.

Chapter 4

Human Societies and the Social Models of Reality

The evolution of human society is involved in a continuous activation of chaotic processes—the same processes present in other aspects of the evolution—subject to specific forces able to move the social system toward an always new, generally unpredictable, situation. Each human group, however large it may be, is a system with many components, open to complex interactions, both internally and with the external environment from which it receives resources and information, apt to create and maintain physical coherences of behavior. The continuous reciprocal interaction among all its own elements and with the outside makes its average world view evolve over time,[10] somehow mediating within it every single mental model of reality (MMR), which anyway maintains its intrinsic individual variability. The result of this interactive process stems from the development of a "social model of reality" (SMR),[11] a "culture,"[12] a specific collective representation of the world, characteristic

---

[10]Obviously by means of much more complex modalities with respect to simpler dissipative systems.

[11]I used the acronym "SMR" for social groups for the first time in Hack et al. (2005).

[12]We share here the concept of culture expressed by Marcia Ascher as "a human group stable in time, which shares language and traditions, together with all the conceptualizations and organizations that characterize our physical and social world" (2007, p. 13).

Myth, Chaos, and Certainty: Notes on Cosmos, Life, and Knowledge
Rosolino Buccheri
Copyright © 2021 Jenny Stanford Publishing Pte. Ltd.
ISBN 978-981-4877-33-6 (Hardcover), 978-1-003-08869-1 (eBook)
www.jennystanford.com

of any specific human group—a representation that contributes, for example, to a consensual definition of the communication and organization rules, of a common language, and of a common methodology for the investigation of nature, including every shared and crystallized stereotypes and prejudices, that will unknowingly become with time a cultural heritage of the group.

It is worth to notice that the concept of the SMR may be applied to any social aspects, from family to school to work activity, within local associations, within our town, region, or the entire nation, and even within a specific international ambit. All these different aspects interfere with each other in an extremely complex way with the aim of contributing to its internal coherence, which is, however, inevitably contaminated by every sort of internal contradictions, some of which have been masterfully described by Luigi Pirandello.[13] As it happens for every single MMR, the stabilization of any SMR, irrespective of the many roles played, is temporary and subject to continuous bifurcation points (new opportunities, exceptional social events, and every other unexpected change within the group) and evolves toward always more advanced social models by means of mutual aggregations with other SMRs, so reducing their number, each shared by larger communities for a longer time.

The validity of the above-expressed thesis appears to be more evident as soon as we look in some detail at the historical development of any social groups, from the political and economic viewpoints, starting from the first settlements of the *Homo sapiens* until the modern megalopolis. A development that has progressed by speeding-ups and turning-backs through encounters and collisions between diverse civilizations, often resulting, from one side in wars and devastations followed by partial or total extermination of conquered populations and appropriation of their territories and assets or, conversely, in the enrichment of the culture of the winning peoples, guided by illuminated sovereigns, who succeeded in integrating their own traditions, so favoring reciprocal homologation. A disordered and unpredictable pathway, then, both over the entire planet and along the centuries, that has traced, since the beginning of human life, a route toward a series of positive improvements in many fields of the knowledge, of the technology, and of the social life organization.

---

[13]See the chapter "Myth–Reason in Hermann Hesse and Luigi Pirandello," Part III.

4.1   The Pathway of Human Societies

As in the case of the single mental models, also the SMRs of any specific human society—mediated over the MMRs of the single components—show differences, more or less substantial with the SMRs of other human groups, in the interpretation of aspects and events open to direct and shared experimentation. These differences are justified by the consideration that diverse societies may use diverse evaluation methods favoring diverse interpretation of same attained data.[14] Concretely, science and technology, the products of the human genius from which we have got today all the extraordinary present knowledge about our physical and biologic world, despite having towed society toward unthinkably high targets of knowledge, were not able, up to now, to eliminate the great variety of elements or events to which single groups do not attribute the same meaning because of the diverse cultural lens with which they are observed. Differences of meanings, these, that may also derive from different life conditions and/or from a different availability of resources in the various regions of the world, whose confrontation may sometimes even aggravate them, as it happens in the today's global context of economic and social crises all over the planet, the origin of a large number of wildfire conflicts, of mixing order and disorder, law and chance.

Concerning the evolution of knowledge, such a mixing becomes a limiting obstacle to the task of reaching an "objectivity" level, able to lead to good control over the oddness of nature; a limit today encountered because of the remarkable growing of our perceived "needs" that cause so many negative effects on the ecologic equilibrium of the planet, and of the destructive ability of our armaments, not speaking of the dissipation of our natural resources due, in turn, to the growing real needs of the dramatically increasing world population and to the lack of equilibrium inherent in the vicious production–consumption circle.

Due to the "chaotic" characterization of the knowledge's evolutionary path—as of any other self-organization process—it is not possible to trace a completely "objective" development of the human civilization from which to extract reliable predictions for the future. Perhaps, to the knowledge's history of a people, we could attach a role comparable to that of myths in ancient societies not yet

[14]See the chapter on *A mixing of unpredictable results in the "project" of living.*

able to write. A role that, as Claude Lévi-Strauss writes, is useful to legitimate a social order and a world view able

> "to explain how the things are through how they have been and find a justification of their present state in a past state; to conceive the future in function, both, of this present and of this past."[15]

It is true, in fact, that myths tell stories of the same kind and with the same purpose as we do today, even if they appear different, because they obey a communicative organization whom ancient peoples believed, as we believe the stories we build today. Stories to which we abandon ourselves with alleged security, despite the observation that every context, even every individual, tells to himself/herself a history different and unique at the aim to gather reasons of hope. Marcia Ascher writes,

> "By examining concepts of other peoples, we are stimulated to consider with higher attention our concepts. In particular, they might induce us to identify forgotten hypotheses. It may happen to find that concepts assumed universal are not, while others considered to come only from us, are actually shared by others."[16]

Then, it is not certain that the present reproduces the past, that the future reproduces the present, and that history—as we conceive it today—surely expresses more objective truths than aspirations; it is only certain that the future differs from the present as the present differs from the past. The interpretation of the historical past in terms of the known present is only hypothetical and leaves great room for any interpretation, especially by different cultures.[17]

At any rate, since the use of writing is thought to have started only around 3400 BC, it is not easy to precisely run over the way that led man to join communities larger than his family, causing him to leave the previous natural habitat where his sustenance had been by hunting animals and picking up fruits from trees. All the same, it is not easy to understand which was the path leading man to elaborate his thought—even considering the appearance of the Paleolithic drawings as a proof of symbolic language—together with

---

[15]Lévi-Strauss (2017, p. 126).

[16]Ascher (2007, p. 15).

[17]The fact that 65 million years ago the dinosaurs' extinction broke a story and started another story all incompatible with the previous one seems to be a trivial confirmation of such a thesis.

the techniques to use in order to facilitate his life in the following Neolithic period. We can only approximately reconstruct the past development of primitive communities on the basis of the few and inaccurate documents obtained from the archeological research. Paolo Scarpi writes,

> "The growth of agriculture and the cattle-breeding are probably at the origin of the formation of stable settlements that, with time, shall give rise to true urban structures. These sedentary arrangements, that look like interruptions of the previous life forms, are the most macroscopic product of the phenomenon known as "Neolithic revolution'. Man becomes a manufacturer of food by means of the rearing starting in the X millennium with sheep and with the cultivation of some cereals and legumes."[18]

In such an evolutionary pathway it is possible that the new urban community organizations contributed to the development of writing, which, in turn, favored abstraction processes, so contributing by a virtuous circle to the social organization and to the evolution of forms of collaboration, both within each single human group and between different groups. Collaboration that, besides the production and conservation of resources, will favor many other diverse activities, among which is the elaboration of technologies useful to satisfy the needs of an always more complex and articulated society. The evolution of these associated life systems, in particular, will give rise to the great civilizations of the east Mediterranean Sea. In Scarpi's view,

> "It is not by chance that the Neolithic revolution especially manifested itself in the Middle Orient and there remaining confined. Only this area coincides with the natural habitat in which tamable animals like sheep, goats, cattle and swine lived together, and with the wild ancestors of cereals and legumes on which was based the Neolithic economy. Egypt did not have the same hydrogeological characteristics and we could think that the use of cultivation methods was the result of a relative diffusion. The region of domestication of animals and of cultivation of cereals seems to be enough limited to the hilly strips located between Palestina, Anatolic plateau, Iraq, Iran, until the declivity of the Hindu Kush."[19]

---

[18]Filoramo et al. (1998, p. 8).
[19]Ibid.

The attention given by anthropologists to the study of the uses and behaviors of social groups that, still isolated in far regions of the world, live till now at a primitive status, together with the knowledge of the history of encounters between people, furnishes us enough data to definitely confirm the thesis of a coaction between chance and necessity to activate the evolutionary path of human societies.

A working method, the last, that allows us to get information on past civilizations, not very different from that of the astronomer who, using modern telescopes, observes the cosmos in order to get information about its past. Both the anthropologist and the astronomer observe things and events by far (in space the anthropologists, in time the astronomers), so remaining exempt from the emotional involvement implicit in the closeness (in place and time) of what they observe and study, even in the impossibility to directly live their condition so as to fully understand them in all their details. As, in fact, the observation of cosmic cataclysms—happened millions years ago and arrived to our knowledge only today because of the limited velocity of light—allows us to look at the past, in the same way, the people not yet able to write, still living in various parts of the world, studied by Lévi-Strauss, offer to us a model to understand the past of humanity. A past that equals 99% of the total duration of the life on Earth.

We discover, with the help of anthropologists, that all those human communities by us considered "backward," that is, lacking of today's scientific and technologic knowledge, have remained stable for a long time until when they have encountered other rival communities, thus starting to change by assimilating, often in a chaotic way, new uses, costumes, and techniques. Lévi-Strauss refers to small groups (10 to 100 individuals), spatially apart by only a few days of walking, with less than a 1% growth rate, practically not increasing with time, thus continuously compensating every person dead with a newborn. This almost stable condition had the advantage to prevent the development of infectious diseases that, instead, started to develop in larger communities, especially when contaminated by other populations. Such small groups (the many thousands already existed and the hundreds still existing in isolated regions of the world) give us, as Lévi-Strauss writes, the means to study human societies at the beginning of their social life, still living in a very primitive condition. A very different situation, this, with respect to what happens in today's towns, where tens or even hundreds of thousands of people live in a much more restricted space but under much better social or health conditions for the availability of resources, for communication technology, etc.

The poor population size and the precise knowledge of natural resources, always available in sufficient quantity, allowed primitive small groups, always in strict contact with nature, the subsistence of all the individuals of the group, even of children and old people who worked only a few hours a day, therefore not contributing to the hard work necessary for the production of food. Such a condition gave them a lot of free time to dedicate to imagination, to formulating religious activities, and to providing a defense against the many dangers of the wild nature in which they lived. The knowledge of the skies with their stable stellar configurations, in particular, was greatly developed so as to constitute the primary instrument for the individuation of paths and moves within the wild territories.

Migrations, with the consequential increased demographic consistency, have radically modified the logistics of people by reducing the free space and, at the same time, by mixing beliefs, traditions, sensibility to natural colors, flavors, and odors.

From the beginning of the nineteenth century, the productive processes by means of not very clean technologies started to pollute water and air, while newborn ideologies, fundamentalisms, and totalitarian views appeared, so "chaotically" modifying the social development toward a new, unpredictable future.

Even with such an unknown perspective, anyway, the winding pathway toward new important changes in the peoples' social models may today come only from innovative ideas elaborated by individuals with great personality, communication skills, and great ideals, attracting new social organization methods. Giovanni Boniolo writes,

"Within a perfectly ordered system, both physical and social, every small perturbation doesn't bring anything particularly significant: the system remains in its 'usual' condition […] In short, a chance fluctuation doesn't change much the pathway of a strongly ordered system. The same happens in a totally *chaotic* system, even if for different reasons […] But there are systems, as it happens in almost the totality of social ones, which cannot be defined nor ordered neither chaotic; systems where chance fluctuations play an important role. It may happen that a perturbation lets rise around it a reinforcement situation coming from other close fluctuations, thus strengthening an already existing fluctuation. In this case, the reinforced fluctuation may lead the system toward a new situation, totally different from the initial one."[20]

New situations, however, are not necessarily positive (it has to be stressed) if they are the result of utopian, too rigid, ideals that enter into our SMR new prejudices, portending social tensions. Nor if such new views are guided by powerful financial entities (very much possible due to the global communication systems today), able to subliminally distribute false updates, apt to stimulate primordial instincts, so "democratically" capturing the attention and consensus of a much larger possible mass of persons, little informed or just kept little informed, and direct them toward the wanted direction.[21]

4.2   Democracy and Migrations[22]

The historical, anthropologic pathway that led man to organize himself in always more numerous groups, with rules of co-existence always more articulated and able to medially meet the needs of all the components of a group, does not deviate—in its general features and taking into account the greater complexity—from the pathways of the other two evolutions, the object of the present discussion. However, because of the need to take into account an always increasing multitude of persons, drastically different for traditions, for experiences, for material, economical, and qualitative interests and knowledge, and for the greater complexity of the related development, this pathway progresses more slowly with respect to the other two, thus reaching less defined goals in the same time. Even so, the fact of having already reached in many countries, especially in the west of the world, the consciousness that a democratic system may better meet the great variety of people's needs with respect to an authoritarian system based on rules dictated by only a restricted number of persons may be considered a good result.

---

[20]Boniolo (2003, pp. 10–11).
[21]The Cambridge Analytica affair is particularly revealing.
[22]These themes have been discussed in *Dialoghi Mediterranei*, n.26, 2017.

A democratic system, aimed to manage large masses of persons who legitimately interpret in a different way "values" theoretically shared (solidarity, justice, etc.), may be seen as a system tending to distribute the consensus over a few items within which the political and administrative offer is somewhat "mediated." Mediation that unavoidably leaves a more or less important percentage of delusions, especially due to the statistical tails of each of the proposed items, but that remains, in most cases, a majority concerning the expectations of every single citizen. To this intrinsic "imperfection" of the democratic system we have to add the fact that its handlers, even if elected democratically (one head, one vote), are never a significant statistical sample of the electing population, with the consequence that the cited "mediation" is necessarily unbalanced primarily toward the interests, conscious or unconscious, of the elected, secondarily to the interests of the electing majority and only thirdly of all the others. Imbalance that generally gives negative results for everybody, sometimes similar to a despotism, if the electing majority, aiming to obtain more favorable measures, economically or closer to its intrinsic interests, fears, and prejudices, did not want or was not able to avoid manipulations. It is worthwhile to remember the famous Giuseppe Giusti's (1809–1850) verse:

> *It is true that the many pull the few,*
> *Provided is in the many wits and virtue,*
> *But, alas, the few pull the many,*
> *If the many restrains inertia or asininity.*

It is not a trivial imperfection, the above, especially if the same population is lacking the wisdom to evaluate the reality with the cognition of cause, so being able to realize the relativity of all viewpoints, to seriously look for confrontation, to understand diversities and accept their legitimacy. Wisdom that may be absent even without being stiffened by those aprioristic beliefs or personal needs that are often hidden ways to vent our own frustrations[23] or, worse, to accumulate visibility, power, and material goods, in all

---

[23]The phenomenon is today amplified by the "democracy of the *clic*" in the web, where the opinion is often an opinion "against," hidden within the mass, and generally aimed to destroy, not to build.

cases alimented by preservation spirit.[24] Features, the last, that for the sake of consensus *tout court* may deviate the course of the public administration, theoretically aimed to serve everybody at their possible best. In view of this, it would be wise to avoid the humdrum crying about politicians and governments—even democratically elected—as if they were made up by aliens, not by humans like us who have voted for them hoping for their specific help.

Such wisdom is due from everybody, including citizens, private associations, entrepreneurs, unions, etc., that should formally and practically concur to meet the regulations decreed by official Institutions or contrast them with true and honest argumentations, not fighting them only on the basis of their own partial view, certainly legitimate but non always objective with respect to their general application.

A particularly evident example of the difficult time lived today by democratic institutions around the world is given by the planetary phenomenon of migrations, causing destabilization of our societies in their usual rhythms and loss of the necessary equilibrium to live in peace within a cohesive setting. Peace felt threatened by a physical and ideological interaction with diametrically different peoples and cultures that becomes always more pressing because of the present cross-movements of large masses of humans, which cause conflicts within the migrants' countries of arrival. Antonino Cusumano, particularly referring to Italy, writes,

> "Immigration, therefore, remains a permanent and diriment object of the political debate, but only as an instrumental matter of rough and continuous electoral campaign, in a Country prone to fight on everything, even on its more recent past, never fully and unanimously shared [...] We should ask ourselves if the integration, other than being a question regarding foreign immigrates living in our cities, is an open problem for us, Italians, who have a weak and uncertain consciousness of our historic-cultural identity, for us Italians who have lost every civic common feeling. As much precarious is our horizon of belonging, all the more opaque our gaze on the 'other' appears and all the more ambiguous reveals our perception of the alterity."[25]

---

[24]The topic was discussed by Buccheri (2017, p. 24). It is quite normal to close oneself in a hedgehog within one's own opinions, refusing a priori any shocking news; it has happened to everybody, even to great personalities like Albert Einstein and Fred Hoyle (see notes 16 and 17 of the Part I).

[25]Cusumano (2017, p. 25).

No doubt that the phenomenon of migrations, within the general context of the knowledge's globalization, has disrupted our living uses, stabilized by decades of economic well-being. A situation stubbornly sought after the disasters caused by the two World Wars in the first half of the nineteenth century. Unfortunately, that well-being has been managed in a disharmonic way by the policies of the physical and economic colonization done by Western governments, who, even if they allowed us to live for decades beyond our intrinsic possibilities, gave rise to a limitation of vital resources for other populations. Today, forced by the adverse circumstances to bitterly realize it, we react in a very confused way and, pushed by the unavoidable presence of prejudices hidden within our mind, unconsciously alter the equilibrium of our MMRs, steering them toward conservation, at any cost, of our well-being, reached by damaging the others in the world. A disequilibrium on which somebody speculates by building conceptual and physical walls, officially to defend our reached economic level threatened by migrations, but actually to stimulate our less noble feelings and, taking instrumental advantage of them, with the scope of easily gaining free and uncontested "democratic" consensus.

The confused fragmentation of opinions discussed here is, in fact, orchestrated by wretched interests, aiming to orient the opinions of the mass, despite the needed exercise of a serious and aware democracy and despite the founding principles of all, universally shared, human cultures. Principles that should warranty the fundamental rights—like the concept of equal opportunities among races, ranks, sexes, attention to the weaker, acceptance of the foreigner, culture of legality—that do not foresee oppression, violence, or a manifestation of superiority.

Exhortations of this kind come very often from especially enlightened personalities. Our Pope Francisco, for example, often invites to build "fraternity bridges" and to "mobilize all energies in order to eliminate separating walls in the world." Also the biologist Francisco Varela has invited to "know what is good and put it into practice," the word "good" being not a moralistic subjective good but the "good for the humanity as a whole."[26]

---

[26]Varela (1996).

4.3 Stereotypes, Prejudices, and the "Common Sense"

As already discussed, any single MMR within a social group, influenced by that of the other members, tends to have a large part in common with them. We have called this common part the SMR, which, in turn, influences every single MMR in an interactive individual–society–individual cycle that leads to that collective, shared representation of the world called common sense, an essential part of every single way to interpret the world and a basis for the construction of the common living within every human group. The common sense, however, implies also the nestling, mostly unconscious, of stereotypes and prejudices, erroneous forms of knowledge that, from one side, allow to synthetize the characteristics of single persons or of social groups and, on the other side, do not reflect always their evolution in time nor their internal differences, thus causing unjustified forms of self-referencing, sometimes also forms of discrimination marked by a presumption of superiority with respect to other groups.

Stereotypes are referred to as cognitive aspects and may be defined as a sort of organized thought, a broad portrait, a schema used for a fast representation of a person or of a social group and formed by a set of distinctive elements—of physical or behavioral kind—considered typical of that group or person that, therefore, will be simplistically described with these features. Stereotypes generally arise from specific examples taken by the knowledge of single persons or communities, known predominantly from literature or by occasional interactions, therefore not catching specific details nor their dynamics. For the way it is built, then, the stereotype ends up by crystallizing the image of a reality, which then becomes part of the common sense, notwithstanding the evident changes of everything with time, as well as of our relations with everything.

The prejudice, defined by Hans George Gadamer as a "predisposition of our aperture toward the world because it looks substantially to the emotional-affective aspect," is a mental behavior derived from the stereotype. It is, practically, a preconceived opinion, not originated by a direct and deepened knowledge of single persons or groups, but simply based on a sensation derived by a fast and occasional contact, from an opinion referred by others, or by an absent will or possibility of deepening. Circumstances, these, that make very difficult to modify the prejudice, even in front of new information able to refute the instinctual, prejudicial, description.

Prejudices are intrinsically irrational, and then very much resistant to any criticism, to any rational confrontation with the reality, and often also resistant with respect to a possible demonstration of their groundlessness. For these reasons, the prejudices nest automatically among the certain and accepted data, even if never directly experimented or rigorously analyzed.

Literature is full of news and studies of stereotypes and prejudices and even furnishes their separation in "light" and "hard."[27] Among the first kind we find both the most classical stereotypes concerning the various people in the world, those not involving dudgeon or offence effects, and, among the "hard" ones, those derived from physical or behavioral flaws occasionally observed and considered stable and those arbitrarily attributed to persons with behavior out of the average or to entire communities with different common sense.[28] In this last category we could include racial discriminations. For them, entire populations, considered "inferior," have been subdued or eliminated in the course of history, often after cruel conquest wars, when their culture, their traditions, and their world view were totally destroyed and their material goods were plundered by the prejudice to possess a superior culture, religion, and world view.[29]

Bloody conflicts born in the light of this belief have often involved social groups of any size, even entire nations, in complex and continuously transforming alliances, with the hidden role to impose their view and their power, together with the need to defend their own values, resources, even existence, at the expenses of other's values, resources, and existence. The result of such conflicts is often dramatic and manifests itself today with millions of peoples fleeing their countries in search of an acceptable existence, possibly in Western countries, risen as a model of good life due to their modern technology and availability of resources. A fleeing, on the other side,

---

[27]Gadamer (1983).

[28]Typical examples of "light" stereotypes are the "precision" of Germans, the coolness of the British, the Italian way to gesticulate, the musicality of Brazilians, the patience of Indians, the Japanese way to go always in large groups, the hardness of Russians, etc.

[29]The very fact that the thirst for power adds almost always to the prejudice of superiority does not change very much these considerations.

that causes a continuous mixing of diverse cultures, which, in turn, let increase—let's hope temporarily—distrust and discrimination. It is very possible that wars and bloodshed caused by economic and value reasons, supported by prejudices of cultural and/or social superiority derived, in turn, by self-referencing and lack of knowledge, will continue for many years or even decades.

Besides those already discussed, there are other (and perhaps much harder) prejudices, generally unknown or not at all considered, because of being implicit in the physical and psychic human structure, to which is connected the unavoidable restriction to perceive and analyze only those phenomena occurring under the terrestrial conditions. The concepts and the language, developed in agreement with the "common sense" bound to such natural restrictions, act as a selection effect everywhere on Earth, therefore losing their validity beyond the strict control of our will and of our organs of sense, where we cannot perform scientifically repeatable and verifiable experiments. These kinds of prejudices affect our common sense, arbitrarily filling the voids of our knowledge of the reality by means of a priori beliefs.

The awareness about the existence of a border to the validity of our common sense within the domain of our daily experience—local, regional, or planetary—destroys our illusion about the universal effectiveness of our reason. Beyond the direct control of the place in which we live, where we cannot even imagine valid observations and experiments, our investigation methods and their consequent results are ineffective or, at least, dubious.

Of course, the difficulties related to unambiguous interpretations of whatever aspect of our real life are present also for our language that, beyond the domain of our direct observations, shows severe limitations. Modern physics, in particular, warns us that certain concepts related to the investigation of the microcosm—then not always directly observable[30]—show in the everyday language important ambiguities of meaning and are only understandable by the use of the powerful but abstract mathematical language,[31] even

---

[30]To a lesser extent the investigation of the macrocosm, observable back in time and only indirectly falsifiable.

[31]For example, those concerning the wave–particle duality, the absence of time in the extreme conditions of matter, the vacuum full of matter appearing and disappearing, particles going back in time or without velocity limitations or not individually existing, and so on.

if some quantum phenomena may also manifest themselves in our dimensions when studied by means of appropriate experiments.

Beyond our dimensions, there may exist an unknown region of the reality not detectable by our "rational system of thought," while we might perhaps think of a remote possibility of access by means of an unconscious perception, which, anyway, we cannot objectify. When it happens, we realize that we must be cautious in defending the idea that what cannot be clearly expressed in words must be wrong or meaningless; perhaps, it cannot be expressed only because we cannot apply to them the current linear logics.

It is unavoidable that our single MMRs and the SMRs of our society may be partially oriented by prejudices nested in some remote place of our mind that manifest themselves sometimes, all of a sudden and unawares. Prejudices exist; we cannot deny or hide them, because, as Jung claimed, despite all our efforts they will continue to exist in us and to manifest out of our control.[32]

4.4 Specialization and Fragmentation, Self-Referencing, and Radicalism

The scientific method is always more used in many fields of knowledge, even with the due differences in precision of its results, especially when applied to those sectors related to the study of very complex phenomena where linear analysis cannot be used. Due to the countless results obtained in the comprehension of the laws ruling our world, so allowing us to decode and control many natural phenomena, the scientific method imposes itself as the rational modality of knowledge par excellence. It is, de facto, the best way known today by man, able to bring us to correct conclusions if starting from experimentally verifiable premises.

Today, all sciences and their connected technologies using the scientific method exercise an always greater influence on our society. However, the very fast increase of the knowledge has caused their subdivision in an always larger number of different specialization sectors, each of them produced by an action of abstraction from the complexity of the global context, able to isolate their intrinsic characteristics in order to have the possibility to study them in

---

[32]Jung (1973).

depth. As a consequence, the whole knowledge has been subdivided into a large number of general areas, like physics, chemistry, sociology, economy, and many others. Each of them, in turn, was subsequently broken down into smaller, more specific, subsections, and so on. Physics, for example, today includes atomic physics, solid-state physics, nuclear physics, thermal physics, astrophysics, etc., and even more subtle subdivisions (astrophysics, for example, can be studied at high energies and at low energies), and furthermore, each of these small sectors may involve, separately, the theoretical knowledge and the experimental methods developed along their historical evolution.

The same is for every other discipline. Given the complexity and deepness reached by any single specialization, it is unavoidable that the comfortable continuous habit in the use of a specific language and on specific concepts and techniques is able to force any scholar to engage all his/her entire life within his/her own sector of competence, without ever coming in contact, in the same deep way, with the ampler context in which his/her sector is merged.

Concerning science, this situation is, according to David Böhm— one of the most important physicists of the past century—the major reason of fragmentation, because it subliminally involves the global set of the large amount of unconscious knowledge that automatically guide our actions, in science as in any other human activity, that is, all those abilities by us learned plus or minus mechanically and that, once acquired, are never forgotten.[33]

Such automatic activities are called by Böhm as "the tacit infrastructure" of ideas and techniques, that is, the set of all that knowledge relative to theories or experimental techniques learned by us day by day and that allow us to be focused on the basic problems of the moment, without having the constant necessity to review all their details. Such a mental behavior, intrinsic to all activities and in all contexts, very often prevents the scholar, as well as the professional of any sector, from keeping constantly in mind the details of his/her own sector with the connections with other disciplines or subdivisions of disciplines and with the general context.

---

[33]Böhm and Peat (2005, p. 20). Simple examples quoted by Böhm are to ride a bicycle, to typewrite, to swim, to replace an electric wire, or to change a seal (for a technician).

Unfortunately, science, as any other activity, is subject to evolution, and all the developments of each sector have often plus or minus relevant consequences on theories, on language, on techniques, experimental or mathematical, of other sectors. As a consequence, the "tacit infrastructure" of knowledge of any sector may become, with time, inappropriate with respect to theories and technical instruments in use and therefore ineffective for the solution of any new appearing problem, without mentioning the possible modifications that follow from such developments within the general context of the knowledge, not always simply recognizable by single scholars.

Actually, it may happen that by studying within a particular sector, one might become aware of the need to extend its context, as it was at the beginning of the past century, when the interest in DNA pushed biologists to use in biology some of the experimental techniques already known in physics, so opening the field of molecular biology, which may now be considered an extension of traditional biology.[34]

At the point in which the scholar starts to face too many contradictions, letting him/her realize the limits of the underlying "tacit structure of knowledge," he/she will finally succeed in questioning theories and hypotheses implicit of his/her sector, and in some cases, he/she will also succeed in redefining the dependence from other sectors. This had already happened, for example, to Max Planck, when he had to "invent" the "light quantum" in order to get out of the "ultraviolet catastrophe" ambiguity, or to Albert Einstein, when he had to postulate the limit to the light velocity in order to avoid the evasiveness of the concept of "ether."

Together with examples of this kind, however, there exists an enormous quantity of cases in which the deepening abstraction act, instead of leading to a new well-defined area connected to the new context, leads to fragmentation.[35] This happens just from what is discussed above, that is, because scholars use their own tacit structure of knowledge in a subliminal way, unconsciously tending

[34]Idem, p. 19.

[35]Böhm and Peat (2005) juxtapose the concept of fragmentation to the image of a clock smashed with a hammer, not producing an appropriate set of subdivisions but arbitrary fragments that do not have any relation with the working of a clock (Idem, p. 16).

to work in the usual way even in front of completely new problems, pushed by the tendency of their mind to base themselves on what is familiar, so resisting against everything that threatens their personal equilibrium and that of the institution for which they work. This trend of the mind manifests itself or by assuming that the unforeseen problem might be solved just by slightly modifying the already used theories or techniques, or by emphasizing the separation between the present sector and other ones so as to work in safety within a limited but known context and without the need to question their own tacit structure of knowledge.[36]

So, unless ambiguities and perceived contradictions are serious enough to cause strong disputes and confrontations, especially at the extralocal level (as it happened at the start of quantum mechanics), the obtained result is the subdivision of the related discipline in different areas weakly connected, both reciprocally and with the ampler contexts. Areas that will become, with time, always more rigidly independent, not anymore valid abstractions of separate fields, as they were born. It follows a form of stubborn fragmentation that will self-sustain by means of the further work done by any other worker in the same sector, guided by the felt objectivity of the made separation from the other areas and from the other scholars, even if the original reason was their ineffective perception.

It has to be stressed that such ineffective procedures might also be reinforced by local problematics of social type because of the strong competition for the hoarding of limited resources and working positions available. The same molecular biology, born as already cited, from the extension in biology of techniques already used in physics, is today fragmented and separated from other fields, such that a molecular biologist has today not much in common with other sectors of the same biology.

The resistance of our "tacit infrastructure" of knowledge is not a prerogative of only science but exists in every field of human life when comfortable and familiar persuasions and beliefs, built along an entire life (especially for old people) or as the result of adhesion to specific ideologies (especially for young people), are threatened by heterodox news that may be dangerous for the stability of our own MMRs. The disturbance of hearing something in contrast

---

[36]Idem, p. 21.

with our own *self-serving bias*, and of being in the need to act in consequence of it, is one of the most important motivations related to the "irrational" choices by us done continuously, resulting in the fragmentation of knowledge and daily actions, different from person to person, very often not coordinated with what is already present within society, fragmentation compared by Böhm to a "broken clock."[37]

The result of all that is a generalized self-referencing of proposals and theories, not always deeply valid in every aspect of social life from single persons or groups who, using the derived confusion of ideas, infiltrate, in an apparently authoritative way, with the aim of proposing exotic solutions, characterized from a generally narrow ambit of knowledge, therefore unbound from the ampler contexts from which they derive. A circumstance, the last, that favors, in turn, an increase of fragmentation, especially when—as it happens in the synthetic exchange of ideas in the web—we use communication means technically advanced but lacking in those precise rules needed to prevent the diffusion of wrong information.

Such a vicious circle, food of self-referencing in all fields of knowledge, is particularly dangerous in the exercise of a country administration and produces a totalitarian and uncompromising radicalism, highlighted today by the growing nationalisms and populisms all over the world. Parodies of utopic cohesions, reciprocally disconnected and conflictual, having nothing in common with the concept of unity of knowledge pursued for centuries by science and by religion, because each of them follows its own "truth," without any solidarity bond, neither within its own community nor at large, in the extralocal context, thus causing a contrasting tangle of "sentences" lacking credibility at all levels, moral and social, only showing an unjustifiable visceral hatred against all that is unknown, despite the reason that would request accurate verification.

---

[37]Idem, p. 19.

Chapter 5

The Conflict between Faith and Science

Due to the intrinsic duality discussed before, the relationship between faith and science causes, since ever and wherever, an intense and controversial condition, always fluctuating between dialogue and conflict. The confrontation becomes particularly pressing and tearing whenever religious faith does not accept the evident achievements of science and whenever science does not accept the limits of the scientific method over questions concerning concepts that go beyond our daily experience.

Religious faith, according to William James, consists

"in the belief that it exists an invisible order and that our supreme Good is an harmonic adaptation to it. The religious attitude of our soul consists in this belief and in this adaptation."[38]

Beyond any single dogmatic and ritual peculiarities, religious belief is exercised everywhere in the world, in a variable way from place to place, but by the vast majority of peoples. Looking more deeply into the question, we may observe that some basic assumptions of each of the many religious beliefs and the following exercise's modalities could largely depend on uses and views elaborated within specific, often millennial, traditions inherited by each local community. Traditions that emerged at a certain stadium of

---

[38]James (1902).

Myth, Chaos, and Certainty: Notes on Cosmos, Life, and Knowledge
Rosolino Buccheri
Copyright © 2021 Jenny Stanford Publishing Pte. Ltd.
ISBN 978-981-4877-33-6 (Hardcover), 978-1-003-08869-1 (eBook)
www.jennystanford.com

the social progress of every people, from the communicative charism of great personalities, whose ideal force was able to train large masses of persons toward epochal ways to approach transcendence, and that were consolidated in the following centuries in their general terms so as to permeate the average culture of future gents. At the same time, anyway, the search for internal self-consistence within each single mental model of reality (MMR) has produced a great variety of fragmentation, of both radical and moderate aspects and perspectives, that has affected the mutual dialogue, generating—especially between different social models of reality (SMRs)—contrasts and conflicts.

5.1 God's Idea and the Gaps of Knowledge

God's idea, as a superior being, includes two different but equally fundamental characteristics, of supernatural type but of different levels. That of the higher level—that almost stably characterizes all deities of history—concerns God's quality to live outside of space and time, so allowing Him to know simultaneously the past, present, and future, while Man is constrained on Earth in a temporary life condition. Together with this supernatural quality that places God beyond human possibilities, God's idea has always contained other characteristics with purely human connotations, even if raised to the superhuman level. Properties, the last, that at variance with those of the transcendence from the physical world, are not the same for the diverse populations on Earth, neither are they stable during human history, with the consequence of producing death and rebirth of every specific God in the course of time. A rebirth, apt to reformulate His representation, in function of the evolution of the knowledge acquired in the meantime by society so as to be more coherent with its new uses and views, which unavoidably permeate the ongoing cultural climate.

As it happens for everybody living within any society, the believers, and particularly the theologians, cannot avoid to perceive such social changes and breathe their atmosphere that, according to the cosmologist and theologian Michał Heller, end with shaping the artistic taste, the evaluation criteria, and the "intellectual preferences of all, believers and not believers, binding them in all everyday choice and behavior, furnishing vital motivations and forming ethical attitudes."[39]

The current image of the world offers always a background from which culture derives its vital force. All humans, even believers and theologians, normally "merged" within the cultural context of the epoch in which they live, make use, implicitly or explicitly, of its current concepts and technical instruments. If the theologian—Heller says—persists in ignoring the results of scientific research,

"he risks to use an obsolete representation of the world, then to limit his pastoral action because people, to day well informed, cannot accept theological truths in conflict with the reality."[40]

Heller's reasoning is always valid; continuous changes in social uses—due, especially today, to technology—may modify with time what is previously shared by tradition, so forcing the theologian to reformulate his/her doctrine.

An idea, that of God, that, by mixing elements of scientific evidence with faith's elements—both unavoidably interlaced within each single conscience and shared within a specific environmental context—proposes a sort of idealization of the human being, transposed beyond his/her space-temporal limits and characterized by general and particular properties, absolute and relative, concerning values taken by a purely human ethic, derived by the concurrence of the two opposite human modalities of knowledge. An idealization arisen because of the real, unsurpassable limit posed by the peculiar human condition—mortal and bound to Earth with its definite physiologic features—for which it is not possible for man to escape toward an objective and absolute knowledge. A limit, this, certified today by the same science with motivations, both physiologic and rational: the first because of the different working of our two cerebral halves, previously discussed, and the other in consideration of the incompleteness of Kurt Gödel's theorems, already cited, that fix the intrinsic limitations of the scientific method within its presuppositions.

Man has always believed to be able to exceed the limits of his terrestrial condition by means of imagination. At the beginning of

---

[39]Heller (2012, pp. 47–48).
[40]Ibid.

the twentieth century, the presumption to have already reached the top of the possible knowledge, induced many people to think to eliminate God from the scene, with the result of finding themselves in front of even more complex questions related to the reached higher levels of knowledge. A condition, this, that can only repropose God's idea, even if with more evolved features and, due to globalization, mediated over an always larger portion of human population. An idea anyway always very complex, where the supernatural properties of omniscience and that of being out of space and time remain almost unaltered, while the "human" features change with time toward forms that depend on the evolving human conditions and on personal and social needs, because of their dependence on all those physical and physiologic peculiarities bound to our mortal terrestrial condition. It is therefore unavoidable that the complex God's idea will continue to depend on our knowledge gaps that can be filled only with an act of faith. An act, anyway, that can only reflect human expectations, of the personal type and bound to the current culture, therefore showing a large range of variations among peoples around the world, even on important concepts like those of "good" on which is based God's idea. Expectations that often, because of this variability, cause harsh conflicts between single persons, between populations, and even between different religions so as to lead to wars and massacres, as the past and recent history continuously tells us.

A restraint to such conflicting differences and to their devastating consequences might be put by processes of cultural homogenization. The direct confrontation between different traditions induced by the advent of the globalized society has certainly become harsher, at least temporarily. Probably with time it will induce, more than in other epochs, the comprehension of the motivations of the observed differences, thus tending to improve the unifying intersubjective datum against the disruptive subjective one. Our hope may be a nascent cultural intersubjective substrate as the final effect of a unifying culture, resulting from a common image of the world that stimulates its comprehension, so creating the premises to look for a reciprocal contact on shared bases, even if somehow hindered by our two opposite forms of thought.

A positive thesis, the last, supported by the present appearance of a conspicuous literary production of philosophical reflections, organization of conventions and conferences at all levels, even of teaching chairs about science and religion, within various universities around the world. Circumstances able to always more stimulate the encounter between believers and nonbelievers, and between believers of different religions.

We cannot avoid finding it very important to see religious organizations all working together, despite the dangerous tendencies to fundamentalism, for a common effort of homogenization toward an ecumenism more aware and respectful of all religious traditions. Working, this, that would be coherent with everyone's own interests in fortifying the common good within the people's conscience.

5.2 Monotheisms Face-to-Face

In every religion, including the monotheistic ones, the religious sense is characterized by a phenomenon of fragmentation at various levels,[41] caused by individual differences of views, in spite of the positive elements of cohesion within few but large groups.[42] The strong and participated cohesion elements often constitute, by the way, a pernicious limit to the development of a constructive dialogue among all the faiths because of prejudices derived by a lack of knowledge of the others' religious culture and/or by hard dogmatic orthodoxy.

Just because they are rooted for centuries within all communities, such uses and beliefs are certainly difficult to contrast without taking the risk of being unpopular or even persecuted as heresies,[43]

---

[41]It is known, for example, that (a) Hebraism contains different viewpoints on the concept of God's omniscience, (b) Christianity is made up of diverse important and organized subdivisions (Catholics, Protestants, Orthodox, Anglicans) and innumerable other differentiations in ritual practice, and (c) within Islam, conflictually cohabit Sciites, Sunnis, and relative subdivisions (Zaydites, Ismailites, and Imamites, plus Hanafites, Malikites, Shafi'ites, and Hanbalites), together with many other "families" and "schools."

[42]The population adhering to the five major religions (Christianity, Islam, Hinduism, Buddhism, Taoism) amounts to about 80% of the world's inhabitants.

[43]To the term "heresy" we attach an idea or an affirmation contrary to a commonly accepted opinion. The concept of heresy may be used in all fields of culture where it exists an orthodoxy of thought and of uses that is overcome by new and strongly contrasting views. We have also to consider the "negative" heresies, those that have produced dramatic involutions and disasters in human history.

as it has happened so many times in history[44] (not only concerning religion).

Such a difficulty, however, has not avoided the rise, within the social context, of personalities of great charisma, great moral tension, and communicative ability, which, challenging with pain and determination the limits posed by the reigning orthodoxy, succeeded to establish new and revolutionary models of life for long periods of time. The revolution activated by Jesus, with his predication supported by an overwhelming and clear moral strength—apt to let him overcome unspeakable suffering and humiliation—is perhaps the most illuminating and sublime example of the refusal of the orthodoxy of customs and prejudices, whatever might be the interpretation from diverse theological or secular positions. A revolution that still fascinates a large multitude—beyond two billion Christians over about eight billion inhabitants on Earth— still keeping almost unchanged its charm after 2000 years. If we would observe this phenomenon in terms of cold statistics, from a Keplerian perspective, it could appear out of normality, given its exceptionality. However, by considering the numbers in this game, that is, the tens of billions of persons who lived since the beginning of human civilization, the probability of such an event is not low, and, in any case, Mehmet's predication, 600 years after that of Jesus with the birth of a new important religious faith, proves it. Also, if we consider the present new orthodoxies of uses, the birth of a new similar event in the future, which will conform to today's higher levels of cultural and social life, is not at all unlikely.

Next, I will focus on monotheisms, inspired by an interesting book by Giorgio Montefoschi and Fiamma Nierenstein, *Only One God, Three Truths*,[45] where the relationships among the three monotheistic religions are analyzed through interviews of their representatives, Claude Geffré for Christians, David Rosen for Jews, and Mustafà Abu Sway for Islam.

---

[44]A significant example within the context of our discussions is given by the story of Akhenaton, the heretic Pharaoh who in 1370 BC tried to impose the monotheistic cult of Aton, without success because of the fierce opposition of Amun's priests, who was faithful to the polytheistic tradition.

[45]The full text of the interview is given in Montefoschi and Nierenstein (2001, pp. 178–189).

To the first question, about the *theological differences between the three monotheistic religions*, the Dominican priest Geffré says that

"Judaism is a religion for the transformation of society toward justice and peace, a messianic religion. Christianity, as we know, is based on the witness of Jesus Christ, a God who is Father, Son and Holy Spirit at the same time. Islam has not any messianic sense of transformation and is the religion of the God's uniqueness, therefore Islam will always refuse Christianity just because it introduces a division within God himself."

The rabbi Rosen, in agreement with the medieval Islamic philosopher al-Bìrunì, is convinced that

"the essence of Christianity is the Trinity while that of Islam is the declaration of the existence of only one God with Mehmet his prophet."

Concerning Judaism, Rosen observes that

"man has to be judged for the way he lives, for his relationship with others and with God. For Jews, theology and ideology are much less important than for the other two religions. From their perspective, the concepts of Trinity and of Jesus' incarnation compromise the 'pure' monotheism idea. God cannot be divided in components or categories."

Concerning Islam, Rosen sees a crucial and insuperable problem in the fact that the Islamic affirm that the Bible is a mystification and that the true sacred text for guiding our lives is the Quran. According to Mustafà,

"The fundamental difference between Islam and Judaism is that the last refuses Mehmet, the Prophet. Between Islam and Christianity, instead, the problem is the concept of Trinity. [Quran] does not speak of Trinity so this concept cannot be accepted; God cannot be like nothing else."

Without entering deeply into the bulk of such affirmations— that should be left to competent theologians—and looking from a Keplerian perspective, it appears evident that any one of the observed differences can only be attributed to pure reasons of faith, for which there is no way of solution from any reasonable viewpoint. Such a condition has been determined, in my opinion, just by the cultural history and by the traditions of the people where the cited religions were born and developed, even with the differences added, later, in the details and in the daily practice. This explanation is almost explicitly confirmed in the second question, *Why are there reciprocal collisions?* To this question Geffré answers that

> "the main source of conflict between Christianity and Hebraism is that the Jews have been considered responsible of the Jesus's death,"

while Rosen's more diplomatic answer focuses more on the fact that every religion affirms to be the one favored by God. From his side, Mustafà deplores that the confrontation between religions is too often based on the crusades' history and on the proclamation of the State of Israel as an important unresolved problem.

It is anyway the third question, *How important is it for your religion to convert the others?* that gives a clearer perception that each of the three representatives considers his religion the "truest" of all. Geffré stresses the "missionary" characterization of Judaism and Christianity, that is, the claim to hold the secret of man's eternal salvation, a characterization denied to Islam. Rosen stresses that God accepts all the righteous of the world, not only if they are Jews; at any rate, Judaism is a style of life, and the passage to another religion is considered sad and unpleasant. Mustafà, finally, does not find errors in Islam, considered by him the most tolerant of the three, because even the "unfaithful"' people are protected by the *shari'a*.

Further to these peculiar affirmations, we also observe from the three representatives' easygoing comments aiming to reduce the reciprocal contrasts with respect to human and social problematics or to greed for power assertions apt to open the door to a tolerant dialogue. From Geffré's viewpoint, many of the reciprocal distrusts between religions might be reduced to a historical dissention between civilizations, in particular between the empire of Byzantium and the religious, political, and, at the same time, military Islamic power. A dissention still existing today because of the Islamic idea that Christianity appears as a religion of the rich and economically developed countries and therefore a too permissive one, while Islam claims to represent the poor populations of the Third World, a religion observing more rigidity both in human relations and in submission to God's law, at variance with respect to the materialism of Western countries, identified with Christianity.

In Rosen's view, it is just human nature to produce such dissentions. Religious motivations were just excuses to justify wars and submission of peoples; religions have always been manipulated to serve other interests. The (tragic) paradox is that armies in reciprocal war would call upon God to defeat the enemy! As if there were two different Gods fighting between them!

Mustafà, as for the previous questions, adopts a more diplomatic line by saying that

> "we should respect the right of existence of everybody, even following his own road and leaving open the door to whoever would enter."

In the light of these interviews, it does not seem singular to affirm that a true reciprocal comprehension and possibility of mediation in a dialogue between different religious faiths (as well as, in general, between different opinions in all other topics) may only be possible by deeply looking inside ourselves in order to understand how much the details of our own faith depend on aprioristic beliefs bound to our own communities—legitimate and respectable in the expression of our own religious beliefs—and therefore hardly to consider as an absolute truth.

5.3 The Opinion of a Theologian

Our knowledge of the universe's structure has today reached a very high level; the mighty mathematical structures of physics and all the observational instruments available today succeed in explaining in very large detail not only the phenomena easily visible with our eyes but also those not directly observable of the atomic and subatomic worlds, without talking of the magnificent manifestations of the cosmos. A knowledge of great quality and quantity, merit of the scientific method made up of logic, mathematics, experimental verification, and interpretation of its results, leading to a coherent image of the world where we live. An image of which everybody makes explicit use, believers and nonbelievers, not only scientists. At this purpose, the scientist-theologian Michał Heller says that "the same Saint Augustine underlined that if a Christian shows his ignorance in science, ridicules the Christian doctrine to the heathens. Nobody would willingly accept ridiculous beliefs."[46]

On the other hand, if it is true that the scientific and technologic culture affects everybody's behavior, it is also true that religiosity— our relationship with the transcendence, to which we refer because of our irrepressible need to ask ourselves any kind of questions on our existence and to give possible answers—is a powerful necessity characterizing all human beings, independently of their faith.

However, if it is true that all of us ask the same questions about the existence and the sense of life, the relative answers may drastically differ. A believer's answer generally refers to a transcending God (with varying features from people to people) to whom to entrust his/her hopes, on Earth and after life, while nonbelievers answer by assuming the self-sufficiency of the cosmos (if they are atheist) or even by thinking to not being able to answer (the agnostics). The religious and the scientific "thought," furthermore, differ on some basic assumptions from which they derive their own conclusions.

Such differences clearly explain why the search for a dialogue often transforms itself into an ideological conflict between single persons or between communities that, not being in a solid equilibrium between the two cerebral tendencies, transform their assumption in a flag for which they fight with determination.

One of the conflictual arguments, on which Heller put his attention, is the one from him ironically indicated as the "God-of-the-gaps theology." It deals with the ideological behavior of many believers that claim to fill the gaps in knowledge with the presence of God or that of many nonbelievers who reject God's idea by presuming that the gaps in knowledge are only temporary. One of the most emblematic examples of "God-of-the-gaps theology" is referred to by Heller when he reports the conclusions of the writer Robert Jastrow, who, next to the scientific debate about the Big Bang, writes in his book *God and the Astronomers*,

"[. . .] for the scientist who has lived by trusting in the power of the reason, the story ends to a nightmare. He climbed the mountains of the ignorance, he is close to conquest the highest top and, soon after he reaches the last rock, he is greeted by a group of theologians who sit there by centuries,"[47]

---

[46]Heller (2012, pp. 47–48).
[47]Jastrow (1978, p. 125); see Chapter 1, Note 16.

so claiming to compare the theory of the Big Bang to the creation of the world by God.

Actually, according to Heller, the Big Bang's discovery cooled the conflict between the biblical theory of creation and the Newtonian theory of an always existing universe, and the theologians hurried to interpret the event as a confirmation about the correctness of the biblical narration. According to Heller, however, the doctrine of the universe's creation transcends the scientific theory of the Big Bang that may, at most, be "consonant" to the creation. Moreover, this theory is a classic one and from the suggestion by Stephen Hawking's works on the evaporation of black holes,[48] our universe could have emerged by means of a quantum process of "tunneling" from a previous condition of physical existence.[49] Such a circumstance does not allow us to associate the Big Bang to the theological concept of creation. The nonbeliever physicist Lee Smolin, in agreement with Heller, writes at this purpose that

> "the overall dominant idea is that the Reason responsible of the coherence of all what we observe cannot be found in the world, but is hidden beyond it,"[50]

perhaps referring to that domain, unknown by the scientific method, that may be individually "guessed"—not in an objectified way—as it happens in meditative introspections and in inspirations, practically by the explication of our "irrational" modality discussed before.

I have widely reiterated that we cannot claim to use the scientific method with the aim of giving judgments of truth on topics that transcend the experimentally verifiable reality. If anything, as beings coming from nature, we could only be able to guess—without the possibility to convincingly explain to others—the possible project from which nature, including ourselves, was born. At this purpose, the physicist Gerald Schröder writes,

> "If the universe is truly the expression of an Idea, our brain could be the only antenna tuned in such a way to capture the information connected to that Idea."[51]

---

[48]Hawking (1993, pp. 133–144).
[49]Heller (2012, p. 9).
[50]Smolin (1998, p. 253).
[51]Schröder (2002, p. 128).

5.4   From a Necessary Dialogue to a Possible Super-Religion

An interesting problem is that of the way to extract from the three monotheistic religions their common fundamental values. A problem whose main difficulty is just to try an agreement between the different dogmatic adhesions and ritual uses crystallized over the centuries, not speaking of the myriads of different ways they are actually lived. Differences not certainly possible to homogenize in a short time, in any case much more beyond our single human lives.

Undoubtedly, the first goal to reach for peaceful cohabitation between different faiths should be that of respectful and sincere reciprocal tolerance, where each of us could keep his/her right to be anchored to his/her traditions without being forced by others' beliefs.

This condition would facilitate the following dialogue's phase, without the danger of a radicalism closed in itself, proud of the "truth" of his/her own faith; pride often amplified by other contingent factors like economic, ideological, or even power matters, always lurking in every individual and social group. All conditions that make it very difficult to reach such a goal, this being the cause of pessimism for the nearby future. The deep self-modifications required in our MMRs are not easily attainable in a short time and in the present conditions, especially in the middle of the clash today observed between different traditions and faiths. A clash that may imply, on the contrary, exasperation of the many, small and great, differences among MMRs and SMRs, almost stable since decennials, and all of a sudden projected in the immediate need for a serious confrontation because of the present epochal migration of entire communities toward places where the professed faiths are all different from those lived before.

An element of optimism for a possible effective mediation in a hopefully not too far future (possibly very much beyond the normal time of a human life) may come from the consideration that, as already discussed, the MMRs and SMRs are stable only temporarily and, even slowly, they are destined to evolve, if nothing else, under the stimulation of occasional social phenomena that, escaping from the "normal" statistics, allow to overcome the current orthodoxies and then urge to move toward a better uniformity of uses and judgments. An equilibrium to which the continuous interaction among peoples of different original cultures and faiths unavoidably has to reach, after having inhabited together for a long time within homogeneous settings and having used the same operative instruments of knowledge.

In an article of his, the Italian Journalist Marco Ventura refers to a project apt to realize a common space for the three monotheistic cults, with a church, a synagogue, and a mosque close to one another, where to break down all barriers "between faiths in the name of an open, tolerant religion," without any reservations on the equality between people of any sex and ethnicities, and

> "on the dialogue between diverse faiths, as a sign, the umpteenth one, of the starting of a new super-religion."[52]

Given the shortness of the article, it is not possible to know what the author actually means for "super-religion," other than the obvious breaking down of known and long-time discussed barriers on themes surely very important but necessarily contingent, simply because they are referred to in the present epoch.

From a Keplerian viewpoint, where no limits of time are foreseen, the present globalization of the knowledge with the epochal transmigrations of peoples (facilitated by the pervasive technology of communications and transportation favoring the common living for long time), could favor the progressive breaking of the walls of intransigence, followed by the equally progressive homologation of traditions, ending with only highlighting the fundamental values to which all cultures obey. A sort of mediation that would culminate in a super-religion, *à la Ventura*, where the mutual distrusts would be overcome, together with any tension due to theological differences and historic conflicts, and where a certain uniformity of evaluations and ritual uses would arise, with a moderated and less conflictual radicalism.

---

[52]Ventura (2016).

Chapter 6

The Search for Equilibrium

The presence of unprovable axioms and the lack of logic solidity of formal systems, previously discussed, together with the unavoidable mixing, within our mental models of reality (MMRs), of the two, reciprocally irreducible knowledge modalities, force our intuition, our creativity, and our rational evaluations to support each other in all endeavors of comprehension and analysis of the external reality. Unfortunately, their participation in such collaboration occurs, as discussed previously, with a clear difference of logic structure, a difference that unavoidably implies the presence of a set of contradictory elements that, in the course of everyday life, have repercussions both in the reasoning and in practical behavior.

This mixing of logics, furthermore, being intrinsic[53] to man's psycho-physical structure and strictly bound to Earth, shows a limitation at a planetary level in our way to evaluate all life situations and the correlated way of expression. This circumstance should alert us for caution when expressing certainties and "objectivity" with reference to events that stay out of any direct verifiable experience.

Given the above, it is also true that man—being produced by nature and then part of it—might potentially keep within his physical and psychic structure traces of his long evolutionary process, with

---

[53]It is intrinsic to our psychophysical structure since it has been defined in the course of the life evolution, starting from its origins on Earth, therefore unavoidably bound to the terrestrial milieu.

Myth, Chaos, and Certainty: Notes on Cosmos, Life, and Knowledge
Rosolino Buccheri
Copyright © 2021 Jenny Stanford Publishing Pte. Ltd.
ISBN 978-981-4877-33-6 (Hardcover), 978-1-003-08869-1 (eBook)
www.jennystanford.com

the consequence of being able to occasionally overcome his direct experience and intuitively capture some aspects of that "implicit order" discussed by David Böhm.

To intercept traces of this kind and keep them under control may be a harbinger of important advancements of the knowledge; in other words, if the consciousness that beyond any direct and verifiable observation may exist an unlimited region of the reality unknown to our rational system of thought is exercised with care, new horizons of knowledge may open, helping us to keep in equilibrium our two knowledge modalities. To have clear in mind that our reason meets a lot of relevant, intrinsic and extrinsic, difficulties when trying to manage our life path in a way free from errors of knowledge is a warning of caution for those who believe in the totalitarian claims of the reason.

6.1   Antinomies' Tension and Control

When discussing the "rational" way with which the crime writers "solve" theoretical cases from themselves imagined, the playwright Friedrich Dürrenmatt writes,

> "Our reason doesn't brighten the world more than the bare minimum. All that is paradoxical fits in the uncertain glow that reigns at its boundaries. We must be careful not to consider these ghosts as if they were something 'in themselves', as if they were outside of the human spirit, or, even worst, don't make the mistake to consider them as an unavoidable error, which could induce us to condemn the world to a sort of stubborn and spiteful moral, whether we would try to impose a perfectly rational view of the things, just because its absolute perfection would constitute its deadly lie and a sign of the worst blindness."[54]

On the side of the religious belief, Mosè Maimònide, aware of the needed integration between physics and metaphysics, already a millennium ago wrote in its *Guide of the Perplexed*,

> "You must know, my son, that as long as you deal with mathematical sciences and with the art of logics, you are in the group of those that turn around the house by looking for the door. [...] If, instead, you have

---

[54]Dürrenmatt (1959, p. 169).

understood physics, then you are within the house and walk in the antechambers; finally, if you have refined the knowledge of physics and understood metaphysics, then you have entered the internal courtyard as a king. This is the level reached by the savants with diverse levels of perfection."[55]

Still today, the theologian and astrophysicist Giuseppe Tanzella-Nitti is firmly convinced that

"[. . .] the scientific-experimental method is not anymore considered as the only rigorous form of rationality, but is flanked more and more often to other forms of rationality: analogic, symbolic, esthetic [. . .] there is an always growing sensibility toward forms of knowledge bound to a Knowledge that cannot be formalized, like tradition, testimony, empathy."[56]

These citations confirm, if it was still needed, that the "scientific" rationality is not exhaustive of our way to understand the world and, in any case, cannot be used to verify any proposition asserted "by faith" (especially concerning religious faith), since they escape from any possibility of empirical verifications. All the same, intuitions and perceptions emerged from our unconscious may not always be used, because of the deficiency of objectivity to which their interpretation is subject, both intrinsically and for the limitations of our language.[57] The problem is insoluble, and also the human condition to possess two complementary and mutually irreducible modalities of thought that come to our rescue along the evolution is not sufficient to assert unquestionable truths. Because of economic problems of synthesis, such a condition forces us to approximate interpretations, constantly submitting us to a tearing dichotomous tension,[58] for which we sometimes favor the Orphic disposition (corresponding to the "nuclear conscience"), while other times we rely on the "Promethean" disposition (corresponding to the "extended conscience"). Mood,

---

[55]Maimònide, parte III, cap. LI.

[56]Tanzella-Nitti (2002, pp. 22–23).

[57]Some ambiguity is always present, even if at different levels, in today's linear language of signs (numbers and letters) as well as in the more complex ones used in the past, like ideograms and hieroglyphs that included images and concepts.

[58]Various types of dichotomies are often described in literature. Let us remember here the loyalty–betrayal dichotomy masterfully described by Arthur Schnitzler in *Traumnovelle* and by Wolfgang Goethe in *Die Wahlverwandtschaften*.

the last, typical of the scientific attitude, that aims at an organized control of our environment and at the empirical verification of every assertion, with a rejection in the absence of demonstration.

It is then understood how it is possible that a disequilibrium between the two dispositions might cause negative psychic phenomena, like the persistence of disturbing memories, their removal, or, in more distressing cases, an almost unilateral development of the conscience, or even the cyclothymic oscillation between the two tendencies.

At any rate, beyond these extreme expressions, every disequilibrium derived from such a dangerous mixing may have other kinds of variable effects that produce contradictions and paradoxes of any seriousness. The harmonization of the two aspects is not easy, and often the conflicts arisen cause incalculable damages to the individual psyche, up to madness. It may then be appropriate, for an effective neutrality principle, a conscious predisposition to a flexibility, able to allow the possibility to enlarge our own MMRs, so avoiding to bar the step to views that frighten us because they are not understood at first sight.

This is, in fact, the lesson we can get by reading about the historic processes that have driven the progress of knowledge. A progress for which we realize that history is a succession of always new acquisitions, able to modify with time uses and traditions, always enlarging the domain of our knowing, so reducing the amount of inconsistencies within our MMRs. I agree with Marina Alfano when she writes that

> "antinomies have imposed themselves as necessary on the evolutionary level, and it cannot be excluded the hypothesis of their harmonic reunification."[59]

Actually, as long as we exercise an aware and fair control that, being "rational" in nature, may only be ascribed to the "extended conscience," the corresponding knowledge processes may, at least partially, converge to an area of compatibility on the pragmatic level, in spite of the antinomy related to the two diverse perspectives, intuitive and rational. Pragmatism that, unfortunately, is often used in a wrong way when tending to satisfy or even amplify our own

---

[59]Alfano and Buccheri (2006).

personal prerogatives to the detriment of others'. All what we can do is an honest and aware search for equilibrium, each of us with the help of our own mixing of intuition and reason. To not exceed in the Orphic disposition would mean to not fall into selfishness and self-referencing; to not exceed in the Promethean disposition would mean to avoid any immoderate rigidity trespassing into utopia.

Within such an awareness we can find the way to check, in quality and quantity, the amount of mixing of the two modalities so as to mitigate their conflicting with the aim to converge on shared values and contribute to a peaceful development of human society. A check that is needed to reduce at least the ideological conflicts—particularly within the religious ambit—by harmonizing the rationality with the positive aspects of the "irrational" modality. Such harmonization is possible within a "complex thought," an extension of the concept of rationality where the reason acknowledge and elaborates the reality by means of a balanced mixing of reason and intuition, without necessarily giving to it an absolute value.[60] A "complex" thought where the concepts of unity and separation may not be reciprocally incompatible if the interpretation of every single fact or event would take the charge to always maintain a strict connection with the general context and its parts, as it should happen, in the social case, for any specializations of the knowledge, born from an aware abstraction from their context with the aim to deepen the details.

To harmonize the connection between the two forms of knowledge, it may be useful to build up a friendly bridge between reason and faith, oriented to the comprehension of the close connections between cultures, tendencies, and expectations. Every one of us, independently of his/her role in the society, is apt to get such harmonization ability, since every one of us—just because of being subject to the dichotomous tension here discussed—is potentially a "philosopher," a thinker,[61] and may have the will and the possibility to face these problems remaining in equilibrium between the two tendencies, to make the best use of the creative tension caused by the antinomy, and acquire a vision as much comprehensive as possible of the world around, able to let him/her exercise with better efficiency all the due tasks.

––––––––––––––––

[60]Alfano and Buccheri (2012, pp. 245–266).

[61]I don't necessarily refer to the professional philosopher, scholar of the philosophic thought.

A "philosophy," then, that is not a profession apt to teach how to live, as it can be offered by a physician or a psychologist with his/her accepted techniques related to particular pathologies, but a philosophy that is the result of a long and thoughtful personal experience, bound to the attention to the various aspects of life, in all physic and cultural ambits, allowing us to capture the image of the world given by the conquests of the thought. A "philosophic" attitude that would act as a bridge among all sectors of knowledge and between the Orphic and Promethean tendencies. There are plenty of examples of personalities of great equilibrium and charisma who were able to bring to syntheses, firmly and judiciously, the common knowledge and related expectations also in the social ambit, without any shortcuts subject to the tyranny of the economy or of personal interest.

6.2 Tolerance as the First Objective

As previously discussed, there are various kinds of gaps and distortions in the elaboration of our experiences, mostly due to prejudices and beliefs, but different in quality and quantity for each of us, according to our role within society. These gaps and distortions are as much hidden and difficult to drive out as they are connected to our physical and psychic structure, making us hardly aware of their presence—an awareness that could help us in the search for their equilibrium with the rational aspects.

Hans George Gadamer underlined the impossibility to ignore prejudices and supported the need for constant confrontation with them and for always calling them into question in order to avoid their crystallization into unmodifiable forms. This means the need for constant attention, because all the unaware a priori assumptions and prejudices unavoidably concur to the process of knowledge and are then transformed with time into consolidated truths.

Gadamer was aware of the necessary emancipation from prejudices and warned us to not take on uncritically all what we get from others and all what tradition hands dawn. A nonperegrine solicitation, the last, that should push everybody to use the meter of tolerance, at least as a mental attitude, useful to critically analyze evaluations and behaviors, ours' and others', in order to purge them, as much as possible, from prejudices. The same, by the way, is suggested by the Declaration of the Principles on the Tolerance of the United Nations, 1995, where we read that tolerance

"is respect, acceptance, and appreciation of the richness and diversity of the cultures of our world, of our forms of expression and of our ways to show our quality of human beings [. . .] an active attitude of all humans heated by the identification of the universal human rights and of the others' fundamental freedoms."[62]

A definition, this, that clearly declares the impossibility that only one man, only one people, only one culture, only one ideology may be the source of absolute truth, but that every single world view may give important contributions to a shared "truth." A statement that should convince everybody to raise tolerance to a "moral obligation to reject all dogmatisms or absolutisms" and to accept that everyone, however different he/she may be in physical aspect, culture, language, behavior, and values, has the right to express his/her own beliefs, only limited by the reciprocal respect.

The term "tolerance," however, should not be only intended as physical or psychologic ability to endure others' uses, language, etc., just in a view of a political or social necessity, but also as the ability to activate and realize a constant and positive confrontation of ideas, in an interculturalism and pacific coexistence view. In other words, tolerance is the ability to assume a healthy, conscious, and pragmatic attitude of deference and listening with respect to all opinions and argumentations, expressed with freedom with the aim to gain and give new knowledge. A kind of attitude opposed to that of a person who, by making acts of faith about ideological or religious hypotheses, closes himself/herself to an extended and comprehensive dialogue, does not accept criticism, and is at most able to reduce the value of tolerance to pure endurance if not to an open attitude of violence.

To support such principles, so contrasting any increase of intolerance that implies a potential thread of conflicts everywhere in the world, UNESCO's Declaration of Principles on Tolerance states that political action must give its support in order to coordinate the answer of the international community to this global challenge, and stimulate to an extensive study on this phenomenon, apt to

---

[62]See the 28th session of the General Conference of the UNESCO Member States, Paris, October 25–November 16, 1995.

undertake effective actions useful to manage it. Support and studies that, to be successful, must have character of rigor and neutrality and therefore base themselves on objective data, obtained by carefully analyzing the specific expressions of ours and others' cultures, in order to extract the best of the common heritage, independently of every physical or ideological difference.

Actually, since we are dealing with the very high complexity phenomena of human behavior, the afore-prescribed exercise is not easy to carry out. We need our best will and engagement in order to produce the awareness necessary to get out of our own self-reference—often manifested as superiority—with incisive listening skills that may put us in a position to discover new realities and new ways of thinking.

A way, this, to amplify the always limited cultural horizon of our MMRs and the "common feeling" of the social group we belong to so as to propose new ideas and solutions for individual and social behavior aimed at a pacific and sustainable development of society. Ideas and solutions that, with time, could become shared and incorporated in the "common feeling."

PART III
INTEGRATIVE AND FINAL
CONSIDERATIONS

Vital Principle

To atoms from quarks, protons, nuclei,
Toward molecules wandering;
Then to all kinds of polymeric forms,
Always in complexity increasing,
Until DNA, replicating cells,
And vegetable, animal, human life.

Simple attraction-repulsion equilibrium?
Or self-organization to Vital Principle due,
In primordial matter already inscribed?

—Rosolino Buccheri, 2018

Premise: The debate about "endophysics" and "exophysics"

Starting from Galileo Galilei and Francis Bacon in the seventeenth century, and along its development during the past centuries, science did not take into account the man–nature interaction, being firm in the belief that nature consisted of a world of objects ruled by an inflexibly causal determinism, characterized by a rigid and unchanging cause–effect relationship that can be observed from the outside by man, who is exempt from its influence. Even Max Planck and Albert Einstein, not yet conscious of any possible conceptual conflicts caused by quantum theories (to which they had given so large contributions), were resolutely certain—the former, that scientific thought

> "aims for causality, actually it is the same as the causal thought, and the final end of every science must be to lead the causal viewpoint till its last consequences,"[1]

and the latter, while writing to Max Born,

> "I am very interested in Bohr's ideas on radiation, but I shouldn't like to be persuaded to leave the strict causality without having fought very differently from the way it has been done so far. The idea that an electron exposed to radiation can freely choose the instant and the direction to make its leap is for me intolerable. If it were true, I should prefer to be a cobbler, or even a gambler, rather than a physicist."[2]

The laws of classical physics were in fact derived by attributing a strict deterministic connection between natural events that not depend by human activity on the base of the belief that man is potentially able to abstract himself from the world to escape from the interactions with the environment and to violate, *à la Descartes*, the dependence between cause and effect.

Such an abstraction implies, in its mathematical formulation, the conflicting condition to find everything trapped in the paradox of an idealized time—without either past or future—in contrast with the human being's perception, which allows him or her to measure the time's current by means of a watch in a really short "psychological present," as if the time flow were only an insignificant

---

[1]Planck (1964, p. 148).
[2]Einstein and Born (1973, p. 98).

human problem from which the rest of the world were excluded. Argumentations that leads us to the notion of *Block Universe*, that is, to the image of a 4D static universe, observable from the outside by man in its spatiotemporal totality.[3]

In the premise of the first part, it was remarked that men are products of nature who live in a continuous reciprocal interaction by receiving energy at a high complexity level, which afterward returns back to nature in a degraded form. During the 1980s, starting from the consideration that it is not possible to think that men are able to analyze the world from the outside without being influenced, a research tendency was developed in order to identify the laws of interaction affecting man's analysis of nature during their co-evolution. This set of hypothetical laws of physics, to be derived with the condition that men must be considered in a continuous and mutual interaction with the external world, were indicated with the term "endophysics" (physics seen from the inside), in contrast with the already working "exophysics," which describes the laws known from classical physics, derived excluding any mutual influence, thus being considered only as an approximation of nature laws. Approximation that, upon the awareness of the enormous values of the scientific progress made during the past centuries, was certainly considered acceptable from a practical aspect, leaving so unresolved the philosophical viewpoint from which it is not acceptable that man is kept separated from the investigated rest of the nature, instead of being seen integrated with it, as if man's observations and actions didn't cause (as they actually do) serious changes in the world observed and, in turn, in the same human being and in the whole community.

Discussions on these themes with the need to elaborate an "endophysics" caused, at the beginning of the 1980s, a debate that lasted intensely for almost 10 years, with a production of a great number of scientific publications but without any practical results. A debate that lately vanished out, especially because of the enormous technical and mathematical difficulties necessary for the requested change in perspective. Difficulties that nowadays appear theoretically justified by the limits to knowledge posed by the circularity of the evolutions mentioned in the first part of this

---

[3]Deutsch (1997, p. 240).

essay. Anyway, the following emersion, both of quantum physics with its dilemmas that cannot be explained in a classical way and of the studies on complex systems with their additional difficulties related to the measurement of precise temporal trajectories, have proposed again the limits within which classical physics is confined, limits that nowadays make impossible to give concrete answers to the derivation of the man–nature interaction laws.

The "exophysical" way of working was formally established within the scientific thought together with the development of modern science in the seventeenth century, thus becoming a peculiarity of classic physics but affecting de facto many aspects of life such to get man accustomed in looking at others—humans, animals, or things—in a detached and less immersive way, often with a defensive attitude, able to increase out of proportion the opinion of ourselves and consequently to tighten individualism and separation.

If only for that, it would be fine to revitalize the concept of "endophysics," not certainly for an impossible mathematical formulation,[4] but at least in order to think deeper about the social life dynamics and to assimilate the idea that the world is not a set of "separated" objects but—as we may assume starting from the doubts posed by modern physics—a web of relations, a sort of superorganism,[5] where man is just a node whose action is reflected on all the other nodes of the web. An idea, in other words, that whatever we think and do is always the result from our interaction with the rest of the world and, in particular, from the intercommunication with the other members of society that, consciously or unconsciously, affects the formation of our beliefs. If we consider that classic physics was founded on the assumption

---

[4]The enormous difficulty is that the mathematics needed to explain "endophysics" should be of a recursive type, where one of the terms to include is just man, involved in the game and describing it as if he were outside.

[5]At the beginning of the 1970s, James Lovelock, being aware of the importance of the mutual interaction between living systems for keeping stable the life on earth, realized that the earth itself may be seen as a sort of superorganism, and called it Gaia (Lovelock, 1979) from the mythological earth's name of Gea. Fritjof Capra, in *The Tao of Physics* (1975) and successively in *The Web of Life* (1996) was perhaps the first to convincingly discuss the universe as a web of relations, thus introducing the concept of "autopoietic system" that was soon after applied and developed to the human organism by Maturana and Varela (1992).

to be an "objective" description of reality, we may understand how dramatic was the crisis of the determinism and separation following the discovery of quantum phenomena. A crisis that has invested also the concept of "objectivity" when it is used to label the results of the scientific work.

Actually, we have to consider that the term "objective" may not necessarily indicate something existing and acting independently from us, but only what derives from a consensus, conscious or unconscious, given to some event or phenomenon, and shared at planetary level as decreed by science—a sort of transposition of the term "subjective" from the single individual to the entire planet.[6] For as much as we strive to hypothesize the presence of only one "objective" truth, we do not have any means allowing us to state it with certainty,[7] considering that even science, as the product of a very large consensus within human society, may found its "truth" only on precise experimental verifications, at planetary conditions.

---

[6]Buccheri (2002).

[7]Outside of our possibilities of observations we could only have intuitions, not necessarily conform to a somehow unknown reality.

Chapter 7

Between Dream and Reality

As Lucio Russo tells us,[8] the modern "scientific thought" was anticipated, a few thousand years ago (between the seventh and sixth centuries BC), by Thales and other ionic philosophers with their attempt to give a rational explanation to the origin of the world. After Thales, other philosophers like Euclid, Archimedes, Eratosthenes, Aristarchus of Samos, and others, in diverse aspects of the knowledge, tried the same.

The tradition tells us that the ancient peoples, fascinated by observing the movements of the stars in the sky, started to think on the world's arrangement. At the beginning, it was expressed through mythic narrations, followed by a critical observation joint with a deep rational analysis also aimed at practical purposes. In any case, the mythic narration and the imagination were not sterile or an end in themselves, but as Karl Popper comments, they prefigured the search for an order in nature. Popper writes,

> "Among the most important myths there are the cosmologic ones, those which explain the structure of the world where we live. The first philosophy and the first science were born in Greece from the critical examination and revision of these cosmologic myths."[9]

---

[8]Russo (2008).
[9]Popper (2001, p. 156).

Myth, Chaos, and Certainty: Notes on Cosmos, Life, and Knowledge
Rosolino Buccheri
Copyright © 2021 Jenny Stanford Publishing Pte. Ltd.
ISBN 978-981-4877-33-6 (Hardcover), 978-1-003-08869-1 (eBook)
www.jennystanford.com

As we see, the knowledge of ancient peoples concerning their relation with the environment consisted of a combination between myth and practical life. As Giuliano Romano writes, "The ancient people were integrators, not differentiators, as we are today,"[10] because they were emotionally involved *in toto* with the ambient. With our present neurophysiologic knowledge, we could say that they used mostly their right cerebral hemisphere. Lévi-Strauss writes,

> "The totalitarian ambition of the primitive mind radically differs from the procedures of the scientific thought. This net difference evidently consists in the fact that such ambition is not realized. Through the scientific thought we can master the nature [. . .] while it is clear that the myth gives them only the illusion to understand the universe, naturally and completely."[11]

The belief in the mythic narrations—as the only true model able to fully explain the origin of the universe and the why of all nature's manifestations—is typical of the just-referred attitude. All the witnesses left by ancient peoples suffer such instinctual modality of knowledge that, being based on cultural bases at all different from those to which our linear investigation methods are today founded, it makes it sometimes very difficult to fully understand its deep meaning. We may only affirm that the "endophysical" dimension, that is, our perception to be part of nature, may be connected to the dimension lived by man at the dawn of civilization, when he had developed a remarkable ability in the (mostly unconscious) evaluation of the interactive situation in which he found himself with nature. A condition which allowed him to realize an optimal global view of what happened around him and to effectually control it with simple but not necessarily aware *feedback* reactions.[12]

With the advent of rationality, we have perhaps lost part of those abilities that would be very useful today in order to reach a deeper awareness of such a fundamental interaction, and that would perhaps restrain our present greedy disposition to use the world together with its structures, living or not living, as a field of wild and indiscriminate predation.

---

[10]Aveni (1994, p. 5), Introduction by Giuliano Romano.
[11]Lévi-Strauss (2002, pp. 31–32).
[12]Alfano and Buccheri (2007, pp. 41–71).

7.1   The Unity of Culture

The concept of One from which everything derives is not disjoint from that of the unity of the culture, the cognitive synthesis to which both myth and sciences aim at: a "Tree of Wisdom: of which Tanzella-Nitti writes in his *Interdisciplinary Dictionary of Science and Faith*,

> "If we see in every science a branch of the tree of the wisdom, then every science appears to us in all its meaning."[13]

A unity which is searched both by *mythos* and by reason, notwithstanding the opposite, irreducible, hostility between these two knowledge modalities whose basic antinomy consists, according to Paolo Gomarasca,[14] in the circumstance that the predominance today of the left hemisphere is bound to problematics of the evolutionary type. In other words, the conflict inherent to the cognitive duality, which sometimes produces results of difficult interpretations, depends very much on the circumstance that in the epoch in which we are living we are consciously bound to a rationality, which, on the other side, is still unconsciously conditioned by *mythos* whose inheritance is anyway always present.

As psychology says, *mythos* keeps us watching inside ourselves and reminding us how much the equipment of our unconscious knowledge is important, especially when it arises unexpectedly to support us in difficult situations. An equipment that manifests it with the intuition, that "irrational" temperament by means of which we are often led instantly and almost unawares to the correct solution of our problems, despite the linear logics of rationality that would propose only a set of possible solutions.

If this is true, and if it is also true that the co-existence of the two modalities has evolved with time tending toward the rationality, then we could explain such a strange combination/contrast of forms of knowledge as a positive resistance, constantly put to reason by myth that puts at service of the evolution its ability to look at things simultaneously from many diverse viewpoints in order to overcome the intrinsic limitation of the linear logics.

Both needs, simultaneously, then: the need for rationalization, born and developed along the human evolution, while our brain

---

[13]Tanzella-Nitti (2002).
[14]Gomarasca (2007).

went specializing itself and activating its specific zones, but also, the urgency to synthesize the many and complex experiences we face along our existence, for which it is not always easy to rationally see their reciprocal influence. A collaborative condition, this, able to satisfy the need to investigate the observed phenomena—of the physical reality as well as of the mind—as deeply as possible, with the means today available to the human's physical structure, so leading our mental constructions to obey a unique complex of rules useful to order all our thoughts in the most efficient way possible, both for intrinsic reason of internal coherence and for reasons of communication and sharing with others.

Unfortunately, within the limiting conditions in which our brain finds itself in the present evolutionary phase, these needs cannot be satisfied in their exceeding complexity. The consequence is a pernicious fragmentation, as before discussed, consisting of the description of a knowledge dispersed into many different disciplines (literature, logics, law, biology, . . .), often without precise interconnections. A culture,[15] therefore, made up of separate sectors because of the brain failure to simultaneously store, comprehend, and retrieve, consciously and globally in all its features, the enormous complexity and quantity of phenomena continuously observed. A culture that, in any case, is the only accessible approximation of the ideal integrated heritage of both forms of knowledge, certainly fragmented but, notwithstanding this limit, necessary to control as much as possible our setting and our vital needs, among which is the need to know the origin of the world where we live and the scope of our presence in it.

In the present limited evolutionary conditions, then, the subdivision of culture into disciplinary sectors has become a necessary step toward the construction of an ordering of all the knowable in view of a dreamt unity of the knowledge. A step that

---

[15]I have found in the literature several different definitions of "culture," from the simplest ones, like "all what is known about every discipline," or "the set of knowing, beliefs or competences," to others more articulated like "the set of spiritual experiences, artistic and scientific realizations matured in a specific place," to the most complex ones, like "the set of intellectual cognitions gained by any person through his studies and experiences, after having elaborated them by means of a deep thinking, such to convert any notions from a simple erudition to a constitutive element of our moral personality, our spirituality, our esthetic taste, and our awareness of ourselves and of our world." Surely a difficult definition!

allows us to enter always deeper into every single discipline but also, paradoxically, that ends with an always netter separation between them, an always larger and more artificially ordered separation, destined to increase with time.

It is a kind of culture that becomes always more fragmented with the deepening of the single disciplines and the continuous arising of always new disciplinary sectors—sometimes only partially interdisciplinary—with the consequent specialization of investigation methodologies and language, especially this last that, because of the always more intense interexchange between different populations, rapidly evolves through a mixing of different typologies of communication. A situation that could let us think to a nebulous future where the continuously increasing quantity of knowledge would cause, paradoxically, an infinite increase of the confusion within a searched organized "order."

We may only hope in an improvement of this condition, consisting of possible future modifications of our DNA that could lead to a human brain more developed and including the *mythos* in more evolved forms in order to allow a more effective collaboration with conscious reason. For the moment, we can only base ourselves on the awareness that all forms of knowledge are able to contain a variable mixing of *mythos* and rationality: from the hardest disciplines—like mathematics, physics, and chemistry where the prevailing rational content marks the precise rules from which to extract precise conclusions from verifiable premises—to others, like literature and poetry, less bound to rigid rules of order but anyway not less incisive concerning communication because of their connotation of intensity of meanings, more synthetically expressed but capable to enter deeply into our comprehension and produce intense emotions full of significance.

It happens that, according to our personal tendencies—if of artistic or scientific prevalence—a concept may be expressed with a variable combination of the two modalities. From an extreme side, with a linear logic language, in a more or less long time and, even if without redundancies, with a quantity of terms as higher as much complex is the concept to express. In the other extreme, with a dense and synthetic language, in a "musical" time full of metaphors and redundancies, of an easy emotional acceptation but not always easy to be fully interpreted. In both cases, all expressions may intrinsically contain elements of incompleteness and difficult comprehensibility with the possibility to be erroneously understood or for the long attention requested or for the need to project to a linear dimension concepts interlaced in a multidimensional way, by using metaphors and allusions, apparently unbound according to the current linear logics. Obviously, the extremes hold only in theoretical bipolar logics: in practice, our every communication touches always the two different typologies with a different percentage for each of us, and it is only our ability to harmoniously integrate them so as to result in a pleasantly comprehensible and acceptable conversation.

In conclusion, however it may seem, dream or reality, the persisting secular search for unity appears paradoxically contrasting with the circumstance of being today in front of processes of fragmentation, sometimes even net separation, of the knowledge in many aspects (cognitive, political, religious, linguistic, . . .), due to the unavoidable factors summarized as follows.[16] First, the uniqueness of the biologic process that characterizes us and that determines our personal predispositions, all different from one another, to interpret the myriad of experiences proposed to us continuously from our setting. Second, the rapid expansion of our knowledge, due to the enormous conquests of science and technology during the past century, much faster than any human possibility to fully store and understand them. Third, the unlikeness of experiences that the course of life proposes to each single person. All factors whose presence catalyzes the diversification of our mental models of reality (MMRs); diversification tending to become stable, helped by the personal unconscious resistance due to the difficulty and to the annoyance to welcome and to integrate ideas and opinions that are incongruous with respect to our own present dispositional behavior.

7.2 The Myth of Unity and the Theory of Everything

The many features of the mythic thought enlightened during tens of passionate studies by historians and anthropologists have disclosed the fact that the ancient mythic narrations could have affected, and still do, our present way of living, evaluating, and acting.

---

[16]See the more detailed discussion in Part II.

One of the most important of these features, of interest for our discussion—a feature that persists within almost all civilizations today—is the atavic, innate, persuasion that the cosmos might be explained by means of a unique entity from which everything emerges and derives.

The birth of the Greek natural philosophy is bound to the concept of unity and to the connected one of perfection. Thales, with the affirmation that "all the things are made up from only one substance," to him attributed by Aristoteles, had put in water the first principle, the origin of everything existing: practically, the first "rational" affirmation about the investigation of the cosmos in human history. After Thales, the unification principle assumed other forms, different from water.

With Pythagoras's arithmetic mysticism, the number became the unifying element of nature, considered symmetric and perfect, and mathematics the key to disclose its secrets. Other unitary ideas of the cosmos concerned the Aristoteles's and Ptolemy's geocentric system with its spheres as additional elements of esthetic perfection, and the following Copernicus's heliocentric system, not forgetting Kepler's unitary cosmos structure published in the *Mysterium Cosmographicum*, resuming the geometric ideas by Pythagoras and Plato.

A dream, an idea, an aspiration, that of unity and perfection, that accompany us since millennia. A wish for which beyond the seeming diversity must exist in the world a unitary principle embracing everything. A wish that may be dated back to the origins of human history and that is present in all the ancient cosmogonies and religions where always a plus or minus transcendental Prime Being dominates, even in the Egyptian, Greek, and Roman polytheisms, where a major God exercises His power toward other minor gods, to Him subject.

To what is connected the tendency, persisting along the centuries, and followed by all cultures, to imagine a truth that overhangs us and of which we all are expressions? Through the knowledge modality here generically denominated with the term "myth," man, as in a dream with opened eyes, throws since ever his own complex life experiences to an ideal scenario characterized by order and harmony, where he may exclude all sorts of difficulties and sufferings so as to live a life full of comforts and satisfactions.[17] To this aim the concept of soul has been invented, for example, to escape the bounds posed by the temporariness of the life on earth so as to survive after death in a supernatural territory, where all sufferings are excluded. Imaginary horizons that, for the fact of transcending the reality, push us to search, with the intellectual means to us available (mainly our intuitions), this fantastic, infinite, and full comprehensive reality, able to conjugate justice with freedom, from everybody always foreboded.

An ideal "order" that we pretend to attribute to the world in view to substitute it to the tumultuous disorder of the obscure forces that escape our control. An order expressed and governed by a sovereign power from us ideally endowed with suitable means, apt to defend the values on which we believe. A hypothetic extraterrestrial mind defined using our terrestrial conceptual abilities, both rational and irrational, and our currently acquired experiences, with the implicit aim to transfer to our living reality the image from us created with such ideal order.

In today's religions, the idea of unity is fundamentally represented by monotheisms, lived by the majority of the world's population, but the search for unity does not belong only to religion. The idea that the cosmos might be explained by a unique theory of everything is present since a long time in science—a theory able to rationally describe all the known, and not yet known, phenomena. A spasmodic search on which converge the efforts of the theoretic physicists of the world is active on this idea since the beginning of the past century.[18]

A search till now without success, mainly because of the impossibility of an experimental verification, as requested by the scientific method, then ending to join to the unverifiable beliefs of all faiths. Paradoxical, a sort of God for those who don't want to talk about God. A dream, an utopic dream, that should furnish the explanation of any kind of phenomenon, a sort of God's mind of logic-mathematic kind that the entire cosmos should obey in all its

---

[17]It recalls the concept of utopia.

[18]For reasons of space, we are not considering here the interesting philosophic speculations on the unity of the cosmos, as derived from the *entanglement* phenomenon studied by quantum physics. An interesting treatment may be found in Böhm (2008).

aspects. A concept that we find, for example, in Roger Penrose, who thinks to mathematics as a unifying essence of the cosmos. Following Pythagoreans, Penrose actually writes,

"One of the most noticeable things in the running of our world is the way it looks based on mathematics at an astonishing level of precision. More we understand the physical world, more we deepen the exploration of the nature's laws, more the physical world appears to dissolve leaving out only mathematics. As more we understand the physics' laws, as more we slide into the world of mathematical concepts."[19]

Today, the *Big Bang* as the origin of the universe and the search of the *Grand Unified Theory* (GUT, the unification of the four fundamental forces) are the basic themes on which modern physics and cosmology work.[20] Presently, all the efforts are concentrated on the theory of the superstrings within the standard model of the elementary particles, considered the natural candidate of the theory of everything so long looked for. Efforts that, together with the attempt to unify quantum mechanics with the relativity theory, have not yet been successful.

In addition, all the studies on complex phenomena, developed since the middle of the past century, have made the situation still worse. This is due to the enormous difficulties inherent to the search of a general theory, able to describe by means of a unique coherent formulation, the totality of the phenomena of our very complex and evolving universe.

In religion as well as in science, then, the idea of unity is now, and could perhaps remain forever, a question of faith that, even if it could have a valid foundation in the atavic man's intuitions, it could remain an utopic idea on the level of its experimental verification and, even if a theory of everything would actually be theoretically formulated, could always have various not demonstrable aspects, with the only result to have obtained an acceptable integration of faith and science (a good result, anyway).

---

[19]Penrose (2000, p. 10).
[20]Concerning the unification of forces, James Clerk Maxwell started in 1864 by unifying electricity and magnetism (actually without including the magnetic monopoles); then, in 1983 the unification of electromagnetism and the weak nuclear force was experimented on at CERN, Geneva.

7.3 Constraints Posed by the Flow of Time

Among the most important hurdles to the harmonization of our two dispositions—the instinctual and the rational—a particular role is assumed by our dependence on the perception of time's flow, unavoidably related to our physical structure, which is bound to life on earth. The human body, as all other living bodies, integrates the natural rhythms of the terrestrial setting (see, for example, circadian rhythms), particularly in the brain, where the temporal phenomena play a fundamental role in the acquisition of knowledge.

Time is, de facto, the most deeply rooted human concept, a belief in which we all are heavily involved for its fundamental influence on our common sense and on our language. All our actions, all our thoughts, are guided by the belief that there will be a future after the present; all what we think, imagine and realize, all what we speak of, is affected by the way we represent and experiment the flow of time,[21] as derived from the elaboration of the external flow of information by our cerebral processes. An elaboration that is harmonized within our organism, therefore implying important perceptual differences between individuals, up to those drastic anomalies called "altered states of the consciousness."

This "subjective" time is the base on which we build our knowledge of nature because of its essential role played as a tool able to manage the quality and quantity of the mixing between rational and relational dispositions with the continuous and unavoidable reciprocal interaction. As such, according to the most recent research in neurophysiology, time is considered a mediator between us and the world,[22] a possible key factor of the modulation between the components of our cerebral structure on which our conscience is based. Time, in conclusion, must be considered the most important among all other influences that filter our deep knowledge of reality.

Do we exactly understand what time is? To forget, to retrieve, to remove; these terms, together with others, recall the flow of time

---

[21]It is a condition that depends on our dimensions. As the theory of relativity tells us, if we could experience the very high velocity of elementary particles, we would observe the distortions of space because of the contraction of distances and the dilation of time; if we would be as large as a galaxy, gravity would modify the measure of time from one point to another of our body and our concept of simultaneity would radically change.

[22]Damasio (2003, pp. 239–240).

and of its perception by man—one of our most mysterious and of almost impossible solution problems, which a lot of literature has dealt with, from Einstein's famous "stubborn and persisting illusion,"[23] successively formalized by the *Block Universe* of David Deutsch previously cited, to the "dynamism in absence of time" to which more recent mathematical-physical studies refer.[24]

In all the cases considered by us as "normal,"[25] an "average" citizen of our times—one who does not suffer from any particular psychological problems and is not subject to drugs—sees the space as a tridimensional reality where he or she can freely move in every direction (with the only and obvious limitations posed by gravity and his or her own physical ability) and conceives time as a unidimensional reality of which he or she can only control the present, while he or she may only remember the past and can only make hypotheses for the future on the basis of his or her knowledge of the present and of the past.[26]

Within these "normal" limitations, to which everyone of us is subject, we can further deepen our observation by highlighting a further level of "normal" variety of perceptions consisting of the presence of differences, both between individuals and in the same person in different psychological conditions. Differences for which time may sometimes seem to flow too much rapidly and sometimes too slowly or even to stop, depending on the experiential situation to which the individual is subject. In addition to these states, as we know, other cases undoubtedly not normal exist—cases quoted in the literature describing situations in which the perception of space and time is lived in a completely distorted way and then connoted as pathologic.

With reference to some recent advancements in physics and mathematics, where dynamic situations in the absence of time have been hypothesized for the prephysical phase of the universe, it has been suggested that time emerged into existence, together with

---

[23]Einstein's opinion communicated by letter to his friend Michele Besso.
[24]Heller (2012, pp. 186–192).
[25]Psychological normality intended in a statistical sense.
[26]The space–time described by Einsteinian relativity let us think of a different reality—a 4D cosmos where space and time are as much interlaced and interdependent as higher is our moving velocity. A condition, however, not belonging to our daily experience, where we move at very low velocities, much less than that of light of 300000 km/s.

space, just after the Big Bang.[27] Time has then guided the evolution, which, presently, is characterized by a cosmological "arrow" directed toward space expansion. However, besides the cosmological arrow, physics talks also of a thermodynamic arrow whose direction goes from order to disorder. However, no one of the two arrows is shown in the equations of the mechanics, where time—in all physics theories, classic, relativistic, quantum—is only represented as a simple parameter, t, made up of an infinite number of instants, all equal to one another, without any internal structure, each one following linearly the other, and where the "now," the precise moment of our actions, is never defined.

In addition to this lack, the presence of the two arrows does not exhaust the complexity of the concept of time; on the contrary it continues to hide the thicker part of Isis's veil, the one under which—according to Eraclitus—nature loves to hide itself. As already discussed at the beginning of Part I concerning dissipative systems, we observe the continuous succession, along the evolution of the universe, of births and deaths of single "complexity islands" within systems interacting reciprocally and with the environment, where the thermodynamic arrow flows toward the opposite direction, from disorder to order, with a gradient that changes all of a sudden, so determining—with the decomposition and the death—the final triumph of disorder and the birth of a new order–disorder cycle. Such "complexity islands," originally discussed by Ludwig Boltzmann[28] and successively studied in detail by Prigogine's thermodynamic of the non-equilibrium, are those in which, through our transitory presence, the Universe investigates and understands Himself.

This difficulty of physics to feature time stimulated Prigogine to consider the necessity to integrate the flow of time within physics. Up to now, however, such integration has revealed to be an uneasy task. The only interesting attempt along this direction refers to the possibility to establish models plus or minus approximated of the behavior of the "subjective" time, a concept accessible to us. The mathematical model worked out by Metod Saniga is able to show a geometric bidimensional map of the three time regions—past, present, and future—together with some of the most famous

---

[27]Heller (2012, pp. 287–290).
[28]Boltzmann (1895, pp. 413–415).

anomalies of the perception of time,[29] without any reference, however, to the human brain and without the possibility to derive from this model new scientific information empirically verifiable/falsifiable.

Susie Wrobel's model of "fractal time" is particularly interesting for the details of its description. In it, the present—the now—is structured as a system of many "nows" stacked one within the other at always minor levels of description, each including the retention, the conscience of the present, and the protentions (memory, now, and anticipation), organized in a fractal way.[30]

All this information, however, is not yet sufficient to reduce the difficulty to reconcile the drastic discrepancy between the parametric time, t, used in physics, with the flowing time of the human relation, notwithstanding the evidence about the perfect working of the first in the description of the (elementary) laws of nature, a description that is at all impossible by any model derived by the analysis of the subjective time. This discrepancy does not allow any identification between the two concepts and, as a consequence, any precise analysis about the fundamental role played by time in human knowledge. We can only try to connect the two concepts by considering the "subjective" time as an (individual) human modality to represent the "objective" time, unknown, that guides the evolution of the universe—and, of course, of our existences— emerged from a (prephysical) phase, whose precise identification and development depend on the used cosmologic system of reference.[31]

From a practical point of view, by considering that the parametric time, t, works so well for the description of the laws of physics, we might accept it as the best-possible representation of the "objective" time, in view of the unanimous consensus within the human community.

These considerations imply that time, whose mysterious and hard peculiarity has always been object of deep discussions, deserves special attention when looking for any possible influence on our choices coming from outside us and suggests that we cannot avoid to include time among the sources of the variability of our individual knowledge. We can just recall the controversy between

---

[29]Saniga (1998, pp. 1071–1086).

[30]Vrobel (2011, pp. 13–33).

[31]Alfano and Buccheri (2009, pp. 100–101).

the old Heraclitus' school of the "real" becoming—followed in the course of time by the somewhat similar philosophic positions by the Pythagoreans, Aristoteles, and Kant—describing time as a real physical event perceived and elaborated by the human mind as well as any other external stimulation, and Parmenides' opposite philosophic option—the "illusive" becoming—later followed by Plotinus and Augustine, according to whom time would be only an idealization of mental contents. In their middle was Paul Ricoeur, who believed in a mixture of these two possibilities. Controversy, the last, that confirms the idea that, accepting or not the existence of a reality external to us, we can never be sure that such hypothesized reality would exactly correspond to the representation we have of it.

The limitations to our knowledge, determined by the necessity to circumscribe our "common sense" within the domain of our dimensions and to the difficulty of objectifying time, together with the intrinsic limitations of our rational system of thought, sum up, thus reinforcing our suspects that a number of regions of reality could be destined to remain at all unknown. At this purpose, we recall Böhm's *explicit and implicit order*[32]—the known and the unknown—concerning the problem if it is really possible to reach (even only theoretically) the mythic full objectivity of the knowledge, since ever pursued by science. As Böhm writes,

"[. . .] every kind of thought, mathematics included, is an abstraction that does not cover the whole reality [. . .] By giving too much emphasis to mathematics, science appear as having lost the amplest context of its view."[33]

7.4   Utopias: Consoling Fantasies?

In the chapter dedicated to the search of a super-religion—longed for by Ventura—the search for an ideal, unified faith was discussed. A search of an uneasy realization because of the great variety of religious traditions and single points of views, together with the difficulty to scratch concepts and behaviors stabilized since centuries.

[32]Böhm (2008).
[33]Böhm and Peat (2005, pp. 9–10).

As in the case of the search for a super-religion, the jarring and ineradicable contrast deriving from the wish for unity—social and political—and the reality of the fragmentation seems to give to unity the taste of a unachievable desire, irreconcilable with our experienced actuality, portending skeins of contradictions that find us constantly engaged to unravel them. One of the ways with which our imagination tries to help us in order to come to terms with them—anyway very often evoked in literature—is to take refuge in utopias, a way that again recalls the concept of unity, but from a personal viewpoint and then bound to the culture and desires of the conceiver.

It happens that we build up a convinced personal opinion upon a dreamed perfection's situation that, inevitably, is declined on the base of our own intellectual and practical experiences, of our listening skills, and of our ability to evaluate and integrate the other's opinions and needs. The emerged mental construction, if strongly and constantly hoped by a personality of high moral sensitivity, perhaps even subject to suffering, leads to the formulation of utopic projects of various kinds, able to obtain visibility and sharing.

History and literature present us a series of utopic proposals for the organization of human society, both of the religious-ecumenical and of the social-political kind. Let's remember Thomas More's *Utopia* or Aldous Huxley's *Island* and others, the oldest of which, Plato's *The Republic*, is located between AD 390 and 360. All literary works, these, speaking of ideal worlds,[34] often considered perfect under the lens of the negative experiences of their authors. Ideal worlds full of obvious educational meanings, rationality, and comfort, but equal for all, like, for example, the city planning and clothing, designed in the way imagined by the utopian, that is, characterized by the most rational simplicity, characteristics that in real societies are, instead, proposed and realized in a variable way so as to account for any personal and social different views.

Each of these and other utopic proposals here quoted have been born out of a complaint toward the present human society burdened by the weight of passivity and dishonesty, and exploited by powerful organizations with the aim of accumulating richness. Complaints that lead to dreaming ideal societies hidden on far territories,

---

[34]The term "utopia" derives from the composition of the Greek words *ū* (not) and *tó-pos* (place), that is, "a not existing place."

isolated from the not loved world and in a condition of a politic and economical autarchy, often of a socialistic type due to the thought absence of private property and to the equal distribution of food and houses. Proposals characterized by the obvious refusal of the war and the equally obvious emphasis on the rationalization of the work, able to support a high level of production for the benefit of the whole population that may dedicate to (theoretical) knowledge a large part of its time.

If we think of the disastrous results achieved by some of these utopic proposals, we realize the clear impossibility of their concretization because of the chaotic situation possibly derived in the long term by the implicit cultural and material differences unavoidably present in each of us. Impossibility certified by vocabularies when defining the term "utopia" as "an ideal aspiration not susceptible to realization."

By the way, is it not at least simplistic thinking to find everybody in agreement about the concept of liberty or of perfection? Which liberty? Which perfection? Would women's communism or the process that leads to the formation of hierarchical classes, as in Campanella's *The City of the Sun*, be immune to all evils or injustices? If the idea of being governed by cultivated persons may always be preferable, as Plato described in his masterpiece *The Republic*, would it be acceptable today that there may exist a specific category of persons, however erudite, immune to the private interests or to any other of the evils that afflict human society today? It appears easier to think that the sincere and naive dream of the utopist, presumed to be objective, may instead have fallen into the dangerous fantasy of a construction where human life, rationalized in all its aspects, is no other than an unaware way to avoid to interact with the crude reality of a varying world, in front of the related responsibilities, and resulting in breeding even greater injustices and sufferings due to the strong control requested in order to well handle such a society. These considerations are correctly highlighted in Hermann Hesse's book *Das Glasperlenspiel*, where the *magister ludi* ends with abandoning the utopic town of Castalia in order to return to live the real life, even if he will have to face the due responsibilities.

As Karl Popper writes, the utopists pretend to have an objective definition of perfection on the basis of which they look forward to a society's radical modification according to principles and rules established not objectively but from their personal viewpoints. Unfortunately, these viewpoints are founded on their adverse experiences that negatively affected their civil and moral sensitivity, pushing them to escape from a too complex and for them uncontrollable reality, till to imagine a hypothetically perfect society by them thought more simple to manage. All that, while forgetting the large differences of view, unavoidable in all human societies, and forgetting, especially, that even the administration's managers are as fallible as all other human beings.

All the above makes vain the pursuit of the "perfection" at any cost and unrealizable any related project because of the contradiction between the need for freedom and the conflicting need for rigid control structures, notwithstanding the sincere dreams aimed to avoid sufferings.

By commenting on the sovereignty theories, Dario Antiseri, in his introduction to Popper's *The Open Society*, says,

> "Nobody is legitimated, by nature, to command over others. [. . .] We have to ask ourselves how can we organize the politic Institutions such to avoid that bad and incompetent ruler make too much damage?"[35]

A comment, this, that implies the importance of democracy as the best, although still imperfect, answer; democracy as the best-possible antidote toward every totalitarianism implicit in all utopias, whose impracticability consists just in the claim to rationalize all aspects of human life, to minutely rule them, even the most private ones, in view of a dreamed theoretical sublime simplicity and perfection.

Let us discuss a specific example of utopia. The Buddhist monk Anthony Elenjimittam, Mahatma Gandhi's disciple known by the name of Bhikshu Ishabodananda, was an Indian philosopher, theologian, writer, and catholic priest of the Dominican Order. His life gives a good example for discussing the contradictions inherent to his ecumenical proposal, probably caused by the sufferings he had during his life, from which the creative construction of his great ideal derived after many years of inner labor. Many aspects of Father Anthony's ecumenical utopia characterize almost all other utopias. When he thinks that

---

[35]Popper and Lorentz (1989, pp. 16–17).

"only the interior evidence, clearly view and experimented by ourselves, may be taken as a criterion of truth,"[36]

we might ask to ourselves whether we do not risk to fall into the radicalism when trying to interpret such "interior evidence" supposed "clearly viewed and experimented" without a safe and wise guide helping to control our limitations and contradictions? Is there not a further risk to worsen our fragmentation and its consequences? Hidden by the light of the ideal longed for, such a contradiction may be revealed in the following affirmation:

"Father Anthony's lesson tells us that anywhere in the world there have been prophets who proclaimed the need that all peoples would join their efforts to relieve humanity's sufferings. [...] The exhortations of these prophets should convince us to realize the Plato' *Republic*, the Thomas More's *Utopia*, the Dante's monarchy, the Campanella's *Town of Sun*, and the *An only world* by Wendell Wilkie."[37]

Let's note that the cited utopists are called "prophets," that is, persons who claim to speak upon divine inspiration. By considering their evident reciprocal contradictions, it is not clear whether such utopias may be really by inspired by God. In any case, in agreement with Popper, it is not easy to believe that the cited utopic projects may really alleviate sufferings, admitted to be able to realize them! The final question posed by Osnato allows us to see, in its full complexity, the practical difficulty to extract the fundamental values, common to all religions. We read,

"Jesus and Gandhi have shown that the force of a single man may become the force of an entire people. The exhortation we got is that to rely on the feeble voice coming from our hearth, and to not fear to testimony for what we have lived and for what we are ready to die."[38]

Father Anthony, of course, refers to the positive voices, those that recall the values of love and solidarity; altogether, we cannot avoid to show up that these voices are generally hidden among the great quantity and variety of voices of any other kinds, coming too from our interior, but not all positive, even if there can be some

---

[36]Ellenjimittam (2001, Cap. 5).
[37]Osnato (2016, p. 48).
[38]Ibid, p. 49.

personal elements of positivity for each of us. To understand such difficulties, it is enough to do some introspection and, for the most striking cases, to study the clinical cases described by the famous psychiatrist James Hillman.

As a matter of fact, one thing is to rely to figures with the moral greatness like Jesus or Gandhi, trying to follow their virtuous path; another thing is that to listen to, *sic et simpliciter*, "the feeble voice which inhabit our hearts" or even "to not fear to testimony for what we have lived and for what we are ready to die,"[39] with the risk to fall into what is called by Hillman "false belief" (when it is only an erroneous belief), which might become the paranoid delirium of the classic psychiatry if it becomes incorrigible, impervious to persuasion,[40] to the reason's logic and to the evidence of senses as the chronicle of our times teaches us!

In conclusion, the rebuilding of an ideal world where all of us might be happy and free from suffering is not practically realizable; the evolution always goes on according to law and chance, guiding our individual MMRs and society traditions. They cannot be modified to our liking and started over, unless using authoritative methods that have nothing of ideally sublime.

Anyway, beyond their impracticability and the diverse ways in which they can be expressed, utopias may have an important role just because they come out as a refusal to the distortions, from moral and physic sufferings, present within the real society, and as a longing for a human society able to alleviate these sufferings. Father Anthony's cosmic ecumenism does not escape this role in view of its action that stimulates the introspective search for fundamental values, of which all of us feel the need. Even more so, our search must be guided by a serious selective discernment of our interior voices in harmony with the common good, helped by an all-out comparison able to enlarge our limits, if possible under the guide of persons of great moral and communicative skills, able to interpret those voices, persons difficult to meet within the chaos of hypocrisy, tendentious propaganda, and cynicism prevailing today.

---

[39]Ibid.
[40]Hillman (2005, p. 18).

Chapter 8

About the More General Meaning of "Myth"

It is time to clarify that the meaning of the term "myth" does not apply only to the narration of the myths of the origin, but it is today accepted in a much broader sense, such as to embrace the multiplicity of meanings acquired over time, and therefore including all forms of speech where we use expression's methodologies that differ from the usual logic-conceptual argumentations. As examples, we may think to the use of unverifiable affirmations inserted at the beginning of a sentence, even within an apophantic speech, when the narration makes use of allegories and metaphors or, when being formally expressed by means of predicative propositions, the sentence falls within the sphere of the semantic speech, as it happens in the presence of extraordinary human feelings or sensations, artistic inspiration or meditation, or historical events lacking safe documents or witnesses. All these cases, besides those referring to the myths of the origins, make clear that the *mythos* is an ineradicable intellectual human attitude with respect to the external reality; a circumstance, this, that poses the myth as the basis of the culture of all peoples, thus assuming today a specific relevance in the study of social sciences.[41]

_____

[41]Buccheri and Buccheri (2005b, pp. 11–27).

Myth, Chaos, and Certainty: Notes on Cosmos, Life, and Knowledge
Rosolino Buccheri
Copyright © 2021 Jenny Stanford Publishing Pte. Ltd.
ISBN 978-981-4877-33-6 (Hardcover), 978-1-003-08869-1 (eBook)
www.jennystanford.com

Watched from this perspective and in the light of the events derived from history and from chronicles, we may notice how the myth, in its most extended meaning, may fill the holes of the rationality in all sectors of knowledge and how the rationality has always benefitted of it, as eloquently explained by Karl Popper when he wrote that the psychoanalytic theories by Freud and Adler were irrefutable but not controllable, even without any doubt of their importance:

> "These theories describe some fact in the manner of myths. They contain interesting psychologic suggestions, but in a form that cannot be subject to control. [. . .] I was aware that these myths might be developed and become controllable [. . .] that from a historic viewpoint, all or almost all, the scientific theories derive from myths."[42]

These considerations led Popper to say that myths may contain important anticipations of the scientific theories notwithstanding the clear difference existing between the scientific and the mythological discourses.

8.1 The Positive Role of Myth along the Development of Sciences

If we look at myth in all its meanings, we realize that it has had a fundamental role in the development of scientific thought, a role able to rehabilitate it on an epistemological level, particularly for all those important aspects related to the mental attitude concerning creativity and intuition. Actually, if it is true that a precise analytic examination of a scientific procedure is essential, it is also true that creativity and intuitions assume an even more fundamental role, that is, that to indicate, since the beginning, the possible operative paths by means of sudden dazzling images emerged from our interior.[43]

This happens in all the works of art that use metaphors and allegoric images, sometimes perceivable only in a subliminal form but able to effectively include more than only one meaning, despite its realization being bound to logic construction rules.

---

[42]Popper (1972, pp. 68–69).
[43]Remember the famous Archimedes' *Eureka*, which suddenly solved the crown's problem proposed by Gerone.

This meeting between myth and science—a sort of "superlogics" able to unify conscience and unconscious, thought and emotion—recalls Leopardi's search for an ultraphilosophy, able "to think poetry without invading its field" or, vice versa, the search for a poetry that "nothing taking away from the truth" could use philosophy without necessarily becoming "reasoning" or defense of a thesis.[44] I seem to be reading Matte Blanco when he uses Pascal's speaking of "the heart's reasons that the reason does not know," and of "the reasons of the reason that do not know the heart's reasons."

Each of us, just because endowed with two distinct knowledge modalities, is a human being with his or her intellect integrated with his or her heart and then he or she cannot think without feeling or feel without thinking, always allowing thought

"to be merged in the reasons of the heart even without understanding them rationally. When he succeeds, he enters into a world that is not foreign to understanding but surely alien to knowledge furnished to us by only the thought."[45]

8.2 Rationality and Myth in Music, Architecture, and Literature

In a comparative study like the present one, where we try to combine the balance between chance and necessity, between intuition and reason, one thing should be reasonably certain, that is, that the first and more fundamental man's intuitions are those deriving from the fact of man himself being nature's product. Upon such certainty I intend to base my further considerations.

If, for example, we look at myth in the ampler sense indicated before, of something that amazes us and influences us subliminally by supplying us sudden intuitions, or unawares leading us to the sudden solution of a problem or to the spontaneous production of an artwork, what more mythic of the concept of beauty may we think of? Why should beauty attract us so much? It is not, perhaps, because man, as a product of nature, is intrinsically sensible to nature of which he is part? The beauty that we so much admire in

---

[44]"[. . .] The reason needs imagination and illusions after having them destroyed," (Leopardi, 1991).
[45]Bria and Oneroso (2015, p. 98).

natural configurations could simply be nothing but the fact that we are connate to nature. Man cannot be insensitive to the nature from which he derives, and therefore he cannot but consider similar (then attracting) all nature's manifestations, even when they are somewhat terrifying, as a stormy sea can be, frightening and attractive at the same time!

Also rigorously rational argumentations, like mathematics— as well as nature's manifestations—have stimulated, along the centuries, great esthetic attraction for the order from them expressed. The pleasantness of music, of the architectonic form of a structure, of a physical or mathematical reasoning, and of everything inspires humans' perception organs, whether they are visual or sound or even abstract, has always been object of attraction by man for the enjoinment he derives from it and to which he since always dedicates attention and study.

The Pythagoreans were interested in music since the sixth century BC; music is today an important matter of study by neurophysiology and physics. The first has confirmed the activation of specific areas of the brain (by the way the same areas activated by sexual orgasm in the case of consonant chords), thus demonstrating the extraordinary physicality of the musical enjoinment that, according to Andrea Frova

"sums up to the motor stimulus of the rhythm, the most obviously corporeal, to which even babies and animals are sensitive."[46]

The second—physics—includes since ever a specific topic for studying the frequencies emitted by musical instruments and for their construction. In both cases, the two aspects are also studied in combination of what concerns the temporal succession of the notes and the chords' harmony, in connection with the mathematical rules and the geometric characteristics of the vibrating strings that underlie their pleasantness. To this purpose, Frova writes that

"the ability of music to induce so high levels of pleasure suggest that the development of this art, based on the attention that already the first humans paid to it as a means of interaction with the environment [. . .] sprang from the mental and physical benefit from it procured."[47]

---

[46]Frova (2006, pp. 113–114).
[47]Ibid.

Music, anyway, is not only a means for biological recreation but also "a product of both our biology and our social interactions,"[48] and it has the characteristic to show itself as

"a necessary and integral dimension of our species, that has played a central role in the evolution of the human mind [...] as all other arts; a means to objectify our senses,"[49]

separated from all other arts only

"in the way in which such objectification happens, so being comparable with all other means of communication based on words and, in particular the various languages or the different dialects of the same language [...] The sense which is on the basis of the arts [...] is not something at all undetermined: rather, it remains *essentially* identical for all arts: but it results more or less differentiated with respect to *accidents*."[50]

In the specific ambit of mathematics, for example, the concept of a "golden section"[51] has always had a role of great relevance for the way the sense of beauty is strongly combined with the logic rigor of this scientific discipline. In our everyday language we use the noun "proportion" in order to identify a relation between things or their parts that appear to us as characterized by esthetic harmony and, for this, attracts our attention. In mathematics, the "golden ratio" (also defined as golden section, golden proportion, or even sometimes "divine proportion"), indicated with the Greek symbol ϕ, is an interesting fusion of esthetic quality and quantitative value, capable of affecting our senses, so making pleasant to our eyes all those objects (of any graphic or architectural nature) that exhibit the harmony of the relationship between their parts, and between each of them and the rest of the external setting, given by the golden proportion. A kind of harmony, the golden ratio, that has always inspired thinkers, artists, and scientists so as to raise this relation

[48]Ian Cross in Peretz and Zatorre (2003, p. 43).
[49]von Balthasar (1995, pp. 13–14).
[50]Ibid.
[51]The ratio a/b of two parts in which a segment $a+b$ is divided (with a greater than b) is called the "golden ratio" if it holds the proportion $(a+b)/a = a/b$, that is, if the ratio between the full segment $(a+b)$ and the major part a equals the ratio between the greater part a and the smaller part b. In this case, the ratio a/b equals the irrational number $\phi = 1.618034\ldots$, with an infinite number of decimal digits.

to a principle of equilibrium and symmetry, both in figurative and in architectonic arts, with the aim to stimulate the attention, a circumstance witnessed by the success of so much literature about this topic.

Mario Livio[52] reminds us, as particular examples, of the façade of the Parthenon inscribed in a rectangle whose sides stay in the ratio $1/\phi$, the spiraling structures of marine shells, and many other objects,[53] structured according to the numerical value of the golden ratio, concluding that the beauty of this proportion is a universal law of nature. The literature is full of references to the "divine proportion" as an instrument able to realize extraordinary artistic creations. Ortoli and Witkowski give us some examples, as that of the Romanian prince Matila Costiescu Ghyka, who, with reference to Pythagoras, in his book *The Golden Section*, asserts that "the laws of Number govern the harmony of both the Cosmos and of Beauty," or that of the physicist Gustav Fechner, who aims to demonstrate as the majority of persons find the golden rectangle more pleasant than any other," or again that of the psychologist Adolf Zeising, who in 1879 supported the idea that "beauty is proportion" because "beauty is the harmony connecting the unity to multiplicity," and finally, that of Marguerite Neveux, an art historian, who, by analyzing many artists' radiographs of paintings and preparatory sketches, had come to understand that the majority of them were based on the proportion $5/8$, a number differing from $1/\phi$ by less than the width of a brush.[54]

Livio claims that the golden proportion is known since the times of Pythagoras, a circumstance suggesting that the reason why ϕ has aroused so much interest, both in the arts and even among the mystics, could just be the similarity between the large and the small, between the microcosm and the macrocosm, as well as our derivation from nature.

The charm we feel in front of the golden ratio seems to show once again the man–nature connection, a connection demonstrating the communication of the human brain with the cosmos of which it is an integral part.

---

[52]Livio (2012).

[53]Other examples are the petals of the roses, the five-points star-shaped arrangement of the seeds of an apple, the branches of the trees and their leaves, the branching of veins and nerves, the proportions of chemical compounds, the geometry of crystals, and many others.

[54]Ortoli and Witkowski (1998, pp. 142–143).

8.3 Myth–Reason in Hermann Hesse and Luigi Pirandello[55]

The encounter between the two human knowledge modalities is present in many literature pieces. Among these, I like to cite, in particular, *Narziss und Goldmund*, one of the masterpieces of German literature, written in 1933 by Hermann Hesse (Nobel Prize of 1946). By re-reading it after more than 30 years from the first time, I get fascinated as before, both for the beauty of Hesse's narrative exhibition and his poetics, always inspired to the identity's discovery and the description typifying his characters. Today more than before, in this tale more than in Hesse's others, and aware of the continuous information furnished by science—generally unknown almost one century ago—I dare to add to the innumerable positive notes addressed to this great author the consonance of his characters' psychology with the way man's duality is explained by neurologists nowadays.

Hesse expresses almost exactly the same when he points out that the world of images, even if it is less conceptual and abstract, is more varied and richer

> "[. . .] in our school days I told you many times that I considered you an artist [. . .] in writing and in reading you had a certain dislike for abstract concepts and you preferred, in speaking, the words and the sounds with amenable poetic qualities, then words with which one can represent something [. . .] Just there, where images cease, philosophy starts [. . .] the world consisted of images for you, of concepts for me. I always told you that you are not made to become a thinker, but I also added that this is not a deficiency and that you, on the other hand, are a dominator in the field of images."[56]

It is not difficult to notice in the characters Narziss und Goldmund the representation of these two diverse human propensities: the first is acute, rational, and devoted to theoretical studies, while the second is totally taken by senses and by the curiosity to dive into the world.

---

[55]This chapter is a re-elaboration of a couple of articles of mine written for Dialoghi Mediterranei in 2016 and 2017 (see the bibliography).
[56]Hesse (1989, pp. 251–252).

"The thinker tries to know and to represent with the logic the essence of the world. He knows that our intellect and its instrument, the logics, are imperfect, as well as an intelligent artist knows that his brushes and chisels might never perfectly express the radiant essence of an angel or of a saint. However, both of them, the thinker and the artist, try in their own way."[57]

As a matter of fact, among the two characters, Narziss is the one aware about their diversity (that's why the name given to him by Hesse) selflessly controlled by him for both.

"Narziss did not believe to Goldmund's vocation for the ascetic life. He had a singular ability to read in men's hearts and, loving, he read with greater clarity. He saw the nature of Goldmund and, even being its opposite, understood it deeply because it was the other half, the lost half."[58]

Narziss knows that every discipline uses its own language and its own specific techniques and that the images are more important than signs for an artistic mind, but shows anyway that the world of concepts controls that of the images. "Think to mathematics," he says.

"Which kind of representations the numbers contain? What the signs *plus* and *minus* represent? None! When you solve an arithmetic or algebraic problem, no representation helps you, you perform a formal task within learned forms of thought [. . .] let me think in peace and judge my thought by its effects, as I will do, judging your art by your works." [. . .] We can, of course, think without any representation! Thought has nothing to do with representations. It is not accomplished in images but in concepts and forms.[59]

The contradictory crush originated by the encounter myth–reason, observable in the opinion about reality through the explication of our MMRs into actions and behavior, is resumed in a wonderful way in Pirandello's literature. In concluding the present chapter, I will cite some examples of inconsistencies and contradictions which characterize our MMRs by referring to the masterful description made by Pirandello. In his works, man wears in life a mask under

_____
[57]Ibid.
[58]Ibid.
[59]Idem, pp. 252, 254–255.

which he takes refuge in order to hide his own identity, intimidated by society's judgment. Such a mask allows him to show only those values and behaviors shared by the community to which he belongs. It is the simulation of a convenient personality that, even eluding the social conventions with the aim to live according to his propensities, is useful to show an appearance that, becoming substance with time,[60] reinforces the illusion to have a personality aligned with that of his own community, notwithstanding the inconsistence between sentences and actions in the long run may be recognized. Alessandro Pizzorno agrees with Pirandello when he writes,

> "If we claim to understand that any person has her own face, and that it is the outcome of a hidden reality, we are in error [. . .] If, instead, [man] is aware to wear a mask [. . .] he will not make the apparently simple gesture of taking it off, thus believing to recover in this way his own authenticity."[61]

The intertwining of the two modalities and the analysis of their equilibrium are the fundamental feature of Pirandello's literature, of which I shall quote only some very significant examples.

The tangle of contradictions is clear in *L'uomo, la bestia, la virtù* (*Man, the Beast, the Virtue*) where Paolino's mask of respectability and the virtue's mask of Mrs. Perella (who got pregnant after an adulterous relationship with Paolino) get tangled up with the mask of the beast, worn by Captain Perella, her husband, who lives together with a lover. The inconsistency reveals itself as soon as Paolino tries, without success, to convince the others that Mrs. Perella is pregnant by her husband.

A twisted attempt at balance is evident in *La patente* (*The Driving Licence*), where Rosario Chiàrchiaro, an ambiguous and calculating man—escaped from everyone and left without work—decides to take advantage of the identity of the evil-eyed man to him assigned by society by asking to have formalized this presumed ability in order to get a job and an income.

The problem of the unknowability of reality emerges in *Così è se vi pare* (*So It Is if You Like*), where Mr. Ponza and Mrs. Frola, his

---

[60]Remotti writes at this purpose that "[. . .] on one side, the substance is already there, in the appearance, and on the other side, the appearance is charged of important and decisive meanings, such to become itself substance" (2017, pp. 121–122).
[61]Pizzorno (2008, pp. 12–13).

mother-in-law, accuse each other of madness when asked about the identity of Ponza's wife. The contrast is concluded with the words of Ponza's wife who says *I am whoever you believe I am.*

An unhappy equilibrium is reached in *Il fu Mattia Pascal* (*The Was Mattia Pascal*), where the protagonist is obliged to follow the rules and the roles imposed by society, even if he got the chance to get free from a difficult situation by allowing him to assume a new identity, apparently more favorable. However, the reality of the new life, even if different, is just as compulsive as the other, and Mattia ends up living unhappily between the life he would like to live and the life that society lets him live.

The same in *Uno, nessuno e centomila* (*One, Nobody, Hundred Thousand*), where Vitangelo Moscarda, between believing to be "one" and aware of being "hundred thousand" for those who know him, surrenders to be "nobody" after becoming aware that the true "I" is not distinguishable from the others.

The analysis in *Il berretto a sonagli* (*The Rattle Cap*) is more articulated. The monologue on "the three chords" by Mr. Ciampa serves to justify the tangle between the "crazy chord," the "serious chord," and the "civil chord" that try to balance each other within every one of us in order to minimize the cohabitation's conflicts, with the unavoidable result to be forced, as puppets, to interpret the part to us assigned on the stage of life, even at the cost of pretending to be mad and remaining isolated.

In his essay *L'umorismo* (*The Humor*), Pirandello explicitly analyzes the contrast between appearance and reality, between the rationality of the common living and the expressions of the soul emerged from the unconscious. One may instinctively laugh by looking at an old lady with dyed hair, greasy with no one knows what horrible pomade, awkwardly raddled, and dressed in clothes of youth. The following meditation, however, lets us perceive the suffering of that woman who perhaps adorned herself in that way just to "keep the love of her husband to herself," so changing the laugh into tenderness and commiseration.[62] It is, according to Pietro Milone, an identity of opposites pervading the different dimension of feeling that enters into the thought and into the language, violating

---

[62]Pirandello (2001, p. 173).

its logic, and therefore an ambivalence that is the translation of a deepest, unconscious, and impossible identity of opposites, able to interpret Pirandello's humor as a composition of fictions, illusions, and simulations, characterized by a bilogic of *matteblanchian* type.

In his interesting foreword to Pirandello's *The Humor*, Milone writes,

> "[. . .] in Freud it exists also a structural conception of the unconscious, as of something that is constitutively different, since in it—and the dreams, for example, demonstrate it—there is a system of representation from which follows a different 'logic'; when these unconscious contents appear to the rational consciousness, it sifts them with the filter of his own logic, in a word: *composes* them."[63]

According to Milone, Pirandello goes beyond such a "composition" and tries to involve man in his unity of thought-emotion, of rational and unconscious sensibleness that thinks and feels in an opposite way. A "beyond" that Milone identifies in Matte Blanco's unconscious, described as characterizing a new and different logic. A logic that violates the principles of Aristotelian logic and that, together with it, constitutes a bilogic, an intertwining between the categories of the unconscious thought and the total absence of categories, homogeneous and indivisible, that thinks and feels in global terms.[64]

---

[63]Idem, LXXVII.
[64]Ibid.

Chapter 9

Where Is Evolution Bringing Us To?

In Part I of the book, the path of three evolutions from their hypothesized start until today has been synthetically described. It is, of course, a contingent datum that cannot be considered final but that allows us to speculate about what we could expect in our later future. It is then a must to conclude our discussion by trying a forecast, taking into account that, as previously said, the high complexity of the problems concerning the development of the evolution of the universe, life, and knowledge, and their chaotic way to proceed, does not allow us to formulate anything for sure. All we can do is to put together, with balance and prudence, what we already know and elaborate it using both cerebral hemispheres, hoping to not go exceedingly far from some realistic future possibilities.

In general, just because the basis of all the considerations done in the course of our discussion concerns the concept of evolution, we cannot stop considering the present moment as the final climax, the stop of any further change. By doing so, we would forget our starting thesis, that is, that we, humans, are beings evolved from inert matter until conscious life, at the point to understand our origin and to analyze the natural phenomena in which we are merged. This means that we could have ahead of us a future of further progress toward directions at all unknown, presumably toward an always greater complexity. All this, of course, in the absence of cataclysms,

Myth, Chaos, and Certainty: Notes on Cosmos, Life, and Knowledge
Rosolino Buccheri
Copyright © 2021 Jenny Stanford Publishing Pte. Ltd.
ISBN 978-981-4877-33-6 (Hardcover), 978-1-003-08869-1 (eBook)
www.jennystanford.com

by ourselves produced or derived from outside earth or in the hypothesis that earth itself would become inhospitable for humanity.

In the best case, the evolution of human beings could continue with the birth of new genetic mutations, followed by the related environmental adaptation. Mutations that could give rise to the birth of new neuronal specialized connections in our brain for a more effective integration of myth and reason and then toward a further improvement of our cognitive abilities. New emerged attitudes, these, that would allow us to more effectively exploit the experiences made along our life, most of which are today forgotten because of the present limits of our brain, as defined by the insufficient storage–retrieval capacity whose effect is their often psychic removal.

9.1   The Open Society and the Mindful Pluralism

Several circumstances, not yet fully understood but able to shape present society's evolutionary path—although with different riskiness from country to country—have eluded the rules of solidarity connection in our society, thus producing a highly conflictual fragmentation, followed by the flowering of indiscriminate pernicious, social-politic self-references, and sometimes even unrealizable utopic proposals, dressed with positive moral purposes. To these proposals Popper opposed, almost 30 years ago, the concept of an "open society," a still valid proposal, although more difficult to achieve today because of the regurgitations of sovereignty occurring despite our past history.

Popper's open society is a democratic society where the focal point is not to be the attribution of the command to a well-defined class of citizens (as in the case of Plato's *The Republic*) but—considering every citizen equally fallible—the realization of always controllable institutions with the real possibility of a loyal and constructive dissent and not too rigid alternatives.[65]

The acceptance of the human fallibility and the assurance of the persevering vigilance by means of opportune institutional

---

[65]A democracy, Popper says, must not be confused with such simplistic formulas like "the government of the majority" or "the people's government," because a majority may also govern in a tyrannical way, and even the people might choose an oppressor, as history teaches (Popper and Lorentz, 1989, p. 17).

instruments makes Popper's open society a reachable objective, not authoritative or closed to differences, not an utopic yearning of unrealizable societies where to take refuge in a dream. Conversely, having clear the characteristics of a real society with its variety of different mental models of reality (MMRs) and its unknowable evolution, Popper's open society may have the necessary stimulating function to push convincingly toward a limitation of the negative tendencies of the commanding power in order to go toward a definite improvement of solidarity, listening, and empathy. I agree with Giovanni Boniolo when he writes in *Il pulpito e la piazza* (*The Pulpit and the Crowd*),

"I propose a way of looking at democracy that somebody could consider elitist. I do not think this is a correct evaluation because everybody could be part of it; it is enough to have the will to study as much as it needs in order to understand what happens and how to participate, it is enough to have the will to be men and women of honor. 'Honor'. An obsolete term, both in the language and in the behavior, a term that intersects various concepts, like reputation, integrity, respect, morality, sincerity, transparence and so on; terms that always more are becoming knick-knacks that adorn the homes of the crafts on duty, those who seem to have forgotten even their sons and grandsons, and that will be, just them, to pay in future the flake of their cunning."[66]

It is difficult to disagree with Boniolo. Competence coming from attention and study, together with a sense of responsibility, may knowingly contribute to solidarity unity in a fertile and interconnected pluralism of ideas and projects, without the reciprocal fighting occurring in a babel of proposals, born from the self-reference caused by fragmentation. When Antonino Cusumano writes,

"The study of diversities within the framework of a temperate relativism remains the best antidote against every expression of ethnocentrism and of identity fundamentalism,"[67]

he imagines a greater investment in the anthropologic culture—integrated with all the specific competences of its components—

---

[66]Boniolo (2011, p. 15).
[67]Cusumano (2016, p. 10).

that could stimulate its awareness, so becoming an antidote (today more necessary than ever) against the extreme and contrasting ideologies of the past century that, after the unspeakable massacres produced by the presumed "racial superiority" and the decades of tensions between a hypothesis of moderate liberalism and a utopic dictatorship of the proletariat, have produced only a liberalism without the necessary rules able to restrain the excesses of individualism and of the accumulation of power inherent to man's fallibility.

9.2   The Future Human Society: A Superorganism or a Babel Tower?

Without a further evolution of our cerebral possibilities, the future history might not reserve us positive surprises.[68] The fragmentation of knowledge, with the balance among the many self-references—due to ignorance, to self-defense, or even to the always larger number of non-interacting knowledge sectors, closed in themselves—could give as a result a sort of "thermal death of culture"[69] and, as its consequence, the dissolution of the human interaction. Today we are not so far from such a condition; only a small fraction of the world population is aware of the effort necessary to transform the pacific "tolerance" into an effective dialogue, together with that to deepen the connection between the various subdivisions of the knowledge.

A fragmentation that somebody would demagogically rise to a positive model of equality, independently from individual competences and culture, therefore interpreted as an absolute freedom to do and to say whatever the most repressed instincts rule, in defiance of any prospect of convergence upon common objective and interests. Demagogy that clearly derives from an interested, grim, and irresponsible calculation aiming to the consensus of a mass of people, expertly "educated" to an illusory concept of equality.

---

[68]For a pessimistic view concerning a further development of life in the cosmos, see Gleiser (2011, pp. 384–393) and the introduction of the book *Novacene. The Coming Age of Hyperintelligence* by James Lovelock.

[69]The reference is to the condition of a gas in a container where the dynamic equilibrium between the various velocities and energies of the particles equals the thermal death of the gas, which loses any informative feature.

Demagogy that ignores (or that does not want to recognize) that we are all different—not all equal—and that such a diversity is not to be considered a limit but an advantage where everyone can contribute with his or her diversity for building up a cohesive society.

A look at the present situation tells clearly that we are today very far from the possibility to quickly reach a useful result for our society, which, instead, may even fall apart in the direction of a chaotic *Babel Tower*, that is, the triumph of the case against the "necessity." A situation apparently in agreement with the second principle of thermodynamics, if we could apply it *tout court* to the mass of persons who populate the planet.

Persons, of course, cannot be assimilated to the simple particles of gas, only subject to thermal motion; they are thinking units whose aware reciprocal interaction may strongly affect the validity of the second principle of thermodynamics; a fact that gives us good hopes for the future evolution. A hope that may become even more consistent if we accept the hypothesis, already proposed before, that life could be already "inscribed" in the matter, thus constituting in itself a circumstance able to originate a total reversal of the present "thermodynamic" situation. It is not strange to think, as it has already happened along history, to the appearance of special personalities—following or not a genetic mutation—able to stimulate such a cohesion able to lead the social system toward a constructive interaction.[70] A new social configuration where the body of the knowledge is more effectively usable and all the differences could positively contribute to building a planetary organization in which every social group, and every person within it, could take on a specific task, flowing into a shared direction, addressed to the survival and progress of the entire human society. A condition that is analogous—at much higher levels of complexity—to that of a human organism where every single cell finds the way to organize itself, specializing its own task in order to contribute to the life of the entire organism.

All we can do in the present situation, to favor a positive evolution oriented to this scope, is to concentrate our own efforts not only in deepening the single activities and awareness but also, and above all, in studying how to establish and reinforce their connections. An

---

[70]Here I have extended the concept expressed by Giovanni Boniolo, as quoted in Part II, Note 20.

activity, this, that—if we exclude the many *everythinglogists* climbed to the fore for pure visibility reasons—is today seriously pursued only by few, sincerely interested, notwithstanding the great difficulties inherent to the need to unify the language in order to make easier the communication. A close connection, then, between all sorts of competences and activities useful for society, by individuals and social groups, notwithstanding the always unavoidable presence of anomalies, due to the possible lack of inclination for collaboration, cause of decompensation and pathologies, exactly as it happens within the human body when in the presence of ill cells or parasitic bacteria.

In conclusion, a way toward a superorganism is possible, in which, in an unforeseen and unforeseeable time, possibly after long periods of self-destructive chaos, the fragmentation and self-referencing are reduced to the minimum amount necessary for the diverse activities be sufficiently integrated so as to be capable to effectively contribute to the superorganism.

Is it fantasy? Utopic optimism? Maybe, but, excluding an everlasting chaos, ending with the self-destruction of humanity even before the sun's death or following some extraordinary cosmic event, the awareness about the possibility to reach such a condition may constitute a qualitative leap and a hope in the today's chaotic situation.[71]

Popper's open society is reminiscent of such a desired superorganism, but only if guided by a thinking head—a democratically elected government that enacts laws only thinking to the progress of the entire country (not only of its voters)—and where the many groups operating in the various activities of society, in a reciprocal osmosis, succeed in maintaining the system in operation along this line, beyond the related personal interests, changing it only in function of their measured effectiveness.

The exigency to build up the great coalitions of states, as the United States and the European Union followed to the common market, was just thought following this line. Unfortunately, the

---

[71]The indiscriminate waste of resources due to the blind consumerism stimulated by commerce and the generalized increase of the population to the record figure of 7 billion in a few decades are not a good omen for what concerns the increased need for food, as well as the indifference to problems derived by the climatic changes, still connected to the policies of commercial expansion.

raising local sovereignties, and the following upheavals wherever in the world, stimulated by the epochal mass emigration, have stopped—let's hope temporarily—the positive process. To resume the journey, it is necessary that the osmosis among the diverse social groups be spread everywhere so as to include all peoples and so reduce the number of individuals pathologically refractory to the system.

As anticipated, the two possible opposite final results of the future evolution are a superorganism or a Babel Tower[72]: either a "normalization," a sort of "thermal death" in accordance with the second principle of thermodynamics, or a great self-organizing process tending toward a higher level society under the stimulation of unpredictable news derived by an increase in complexity. Opposite directions, both anyway possible due to the always more diffuse migratory phenomenon and the always more pervasive communication and transportation technology that favor a mixing of uses and knowledge, a harbinger of dialogue as well as of self-referencing. Migratory phenomenon, by the way, that will probably last for a long time and that cannot be stopped anymore.

When the mixing of people will reach its maximum and no further emigration, capable to change radically established uses, will be possible, the process of international homologation of uses, language, and culture will end with the birth of the supernation earth.[73] It will certainly take a lot of time, even hundreds or thousands of years, and we who write and read about it now will not anymore be here to verify the result.

Coming back to the comparison between the three evolutions characterized by the chance-necessity, law-fortuity, encounter, we may pose to ourselves the question about their future path, today unknown: will it have an epilogue or will the evolution last forever? The second option will convict us to an eternal pursuit between what we know and the underlying reality. The first, instead, might

---

[72]Even for the evolution of the cosmos we have, approximately, the same two opposite possibilities of conclusion: either the indefinite expansion of the universe, leading to an infinite number of non-interacting black holes, or a *Big Crunch*, a coming back of the expansion (according to the existing amount of total mass) until the reunion of all the mass into an enormous nucleus, which could, in principle, lead to a new Big Bang (Hawking, 1993).

[73]A supernation ready to compete with possible other extraterrestrial supernations, so starting a new cycle at cosmic level.

indicate the ideal reach of the evolutionary limit, the end of the flow of time.

9.3   A Possible Future for Human Beings

The need to serially order our experiences, in accordance with the rooted rationality, may have determined our present choice to live into a precise time direction from past to future. Along this path, the evolution of life had led to the present human being—its present climax—with a cerebral configuration represented by a bilogic leaning on two collaborating cerebral hemispheres. A configuration apt to help the human organism in facing the complexity of everyday life in the limiting physical conditions of the external environment (climate, gravity, temperature, radiations, etc.) and of our organic setting (sight, hearing, smell, weight, breathing, etc.)—a configuration not yet able to take full advantage of all the experiences and knowledge acquired along life that binds us to distribute our knowledge among several different sectors.

In these present limitations, the way in which *mythos* and *logos* are today integrated allows us to reduce the confusion derived by the impossibility to take simultaneously into account all what we experience. This kind of organization may let us think that our brain might acquire, by further evolution, new abilities beyond those today "normally" expressed. After all, even today the information's elaboration ability is not the same for every one of us, as it shows the rise along the history of so many exceptional personalities with much greater intellectual possibilities.[74] In any case, considering that the collaboration between two distinct cerebral hemispheres, irreducible to each other, is generally not sufficient to deeply face the enormous quantity of connections among the various experiences, this difficulty and the following effort to try to overcome it could let rise opportune genetic mutations and related adjustments to better face the challenge of complexity.

The presence in our brain of specialized zones of information's elaboration, highlighted with always greater evidence by biologic researches, encourages us. We have already a lot of experience

---

[74]I refer to those great personalities of all cultures and of all aspects of civil life who have influenced the evolution of society, even if sometimes negatively.

with nature's power; its extraordinary and ingenious way to evolve toward always more complex forms is the hope of mankind. Basing upon this hope, we can think at possible future mutations able to lead toward a higher effectiveness in the integration of *mythos* and reason so as to let us increase our perception of time in a middle and more controllable position between the reason's slavish seriality and the confused time indistinguishability of the mythic knowledge.

Along this line of thought, the already actualized emergence within the left hemisphere of our brain of areas specialized in rational thought, in addition to those areas of the right hemisphere dominated by *mythos*, since always based upon elaborations of instinctive type, could be intended as an evolutionary step in the direction discussed before, a further means devised by nature in order to proceed toward a better understanding of the world, not possible thousands of years ago only by myth, but also not yet possible today with the help of rationality at its present level.

The main image of this book, with its question marks, refers to the circularity of the evolutions on the basis of the limited knowledge today available, but nobody might ensure the perennial presence of such limitations beyond which the development of human intelligence is destined to stop its evolution, without resorting to unprovable dogmatic positions. All the same, nobody can state, without evidence in the hand (as Gleiser does), that in some very far part of the cosmos there could be places where life has already developed, or will develop, until limits today inconceivable to us.

Wagering on the hypothesis of a possible further evolution, it is not said that in a not too far future, a new sentient being could not emerge, at all different from us today, both for the shape and for a physical body more effective in moving (even between planets or even between stellar systems) and for a more effective cerebral structure in the acquisition and elaboration of information, therefore more capable to experiment and deeply investigate the external reality.[75]

---

[75]Let's recall Laplace's dream expressed in his *Essai philosophique sur les probabilités* (1814): "[. . .] An intellect which at a certain moment would know all forces that set nature in motion, and all positions of all items of which nature is composed, if this intellect were also vast enough to submit these data to analysis, it would embrace in a single formula the movements of the greatest bodies of the universe and those of the tiniest atom; for such an intellect nothing would be uncertain and the future just like the past would be present before its eyes."

Personally, I am unmotivated to support the idea that evolution has already arrived to its climax; I believe, instead, it reasonable that the future millennia could reserve to us unsuspected news, even if we who are writing and reading will never know. Moreover, if life were inscribed in the matter as here hypothesized, and if for some unpredictable and dramatic reason the life on our planet would suddenly be extinguished, nothing forbids that the process of self-organization discussed before would not come to completion in some other region of the cosmos, with the emergence of an organism of superior level with respect to the human being as we know today. The developments in the search for extraterrestrial civilizations could give us some insights in the near future.

Afterword

Whatever is found in "practice" must lie within the scope of the metaphysical description. When the description fails to include the "practice" the metaphysics is inadequate and requires revision.

—Alfred North Whitehead

Emergence, Life, and Complexity

1. Overture

When Rosolino Buccheri asked me to write an Afterword to his book, I was very pleased and honored. Indeed, as a scientist with a broad range of philosophical interests, he's able to produce theoretical works of great value and importance. Among these, *Myth, Chaos, and Certainty: Notes on Cosmos, Life, and Knowledge* is surely a representative example of this kind of work.[76]

It is enough to browse through its pages to recognize the careful design of a complex conceptual architecture. Having a considerable knowledge of different fields of science (physical, social, biological,

[76]I would like to thank Rosolino Buccheri, Yuri Di Liberto and Roberto Lo Presti for helping me with their precious remarks during the writing of this text.

Myth, Chaos, and Certainty: Notes on Cosmos, Life, and Knowledge
Rosolino Buccheri
Copyright © 2021 Jenny Stanford Publishing Pte. Ltd.
ISBN 978-981-4877-33-6 (Hardcover), 978-1-003-08869-1 (eBook)
www.jennystanford.com

etc.), the author has developed a transdisciplinary framework in which different issues and styles of explanation interact with one another.

As the reader will have noticed, the names of writers like Pirandello, Hesse, and Joyce get together with those of audio-psycho-phonologists like Tomatis or psychoanalysts like Matte Blanco. Furthermore, we see a wide variety of topics, including (1) the relationship between *mythos* and *logos*, (2) the fascinating debate between endophysics and exophysics, (3) the social models of reality, and so on. In short, Buccheri gives us a strong heuristic tool for understanding the different types of interactions underlying

> "the many-levelled structure of the world of which humanity is an evolved constituent."[77]

The concept of emergence is claimed to play a pivotal role in the general economy of the book. First, because, according to the author, such a concept represents a source of insights useful to provide an explanatory framework for making sense of the world's complexity. Second, because "the debate about the 'emergence' phenomenon is far from being definitely closed."[78] This last remark needs some comments.

2.   Emergence and Two Types of Constructivism

Buccheri is right when he asserts that emergence is a complex and intricate concept to grasp. As many scholars of different disciplines pointed out, "Defining emergence is not easy because it is not a monolithic term."[79] In this sense, the central position attained by emergence in the discourse of Buccheri is not the side effect of a "naïve vision" but rather the fruit of a refined reasoning developed in the fields of mathematics and natural sciences.

The conceptual path leading the Italian physicist to adopt an *emergentist* point of view is justified by two primary considerations: one concerning the presence of *nonlinearity* effects in several

---

[77]Deane-Drummond et al. (2003, p. 134).
[78]See Part I, "Premise."
[79]Rivera (2010, p. 143).

systems of natural and human world and the other concerning the epistemological necessity of

"a general theory [. . .] able to indicate the necessary and sufficient conditions for the appearance of processes [. . .] irreducible to the laws which rule their elementary constituents."[80]

These two formal aspects, that is, *nonlinearity* and *qualitative discontinuity*, are the main ingredients of the emergence concept worked out by the author. In particular, the criticism addressed to the notion of "elementary particle" clarifies, in my opinion, the kind of emergentism advocated by him. As is well known, many varieties of emergence seem to presuppose a metaphysics of "elementary particles" that is strongly compromised by latent reductionist ontologies. This is the case, for instance, of the so-called L-systems.[81] The patterns generated by rewriting the rules that compose such algorithmic devices globally show a fractal-like structure that is absent at the level of the basic units of computation. So, they are exemplary of the way an emergent structure is supposed to work. But, as I said, they actually represent only a reductionist variant of emergence. There are at least two features in an L-system that permit what might seem an oxymoron. First is the fact that it needs of finite sets of symbols to perform its computations; second is the fact that it lacks appropriate descriptions of top-down constraints that usually characterize the nomic profile of an emergent structure. One could say, therefore, that L-systems are algorithmic instances of a specific form of constructivism. I refer to what Yves-Marie Visetti calls an "assembly constructivism" (*constructivisme assembleur*), which is a particular approach to emergence in which

"il y a [. . .], nécessairement, des éléments primitifs (non construits) dont l'individualité ne se modifie pas en participant à la construction, et subsiste inaltérée dans la trace des procès."[82]

Now, it is this "vision of emergence" that Buccheri refuses to accept. Being a very deserving connoisseur of quantum mechanics, the author knows that

---

[80]See Part I, "Premise."
[81]See Lyndenmayer (1971), Petitot (1979), and Prusinkiewicz (1995). For more details about L-systems, see the appendix.
[82]Visetti (2004, p. 232).

"truly elementary particles, [. . .] in the sense given by Greek atomists, do not exist, and the wave-particle duality is a probative confirmation of it."[83]

Then, to which kind of emergence does the author adhere? From the analysis of Buccheri's essay, it seems that he follows a variant of emergentism that presents all the features of the so-called "genetic emergence" (*émergence génétique*).[84] In this epistemological framework, "particles [. . .] have no substantial existence [. . .] everything is a process."[85] More precisely, everything is made of entities

"qui ne préexistent pas toujours à leur mise en relation, et qui doivent leur identité, leur stabilité, et avant tout leur placement, aux organisations qui les produisent."[86]

Hence, in the perspective defended by Buccheri, emergence is more properly *co-emergence*, which is a process in which

"la construction [. . .] remanie toujours ses constituants (et) [. . .] où le tout et les parties co-adviennent dans un processus de codéterminations croisées."[87]

Probably, research fields like dynamical continuous systems theory constitute an ideal framework to grasp the formal properties of this kind of emergence. Certainly, that subfield known as *dynamical autopoietic systems theory*[88] is, for Buccheri, a point of reference of great interest. But it seems to me that the author prefers to investigate some applicative aspects of the emergence rather than to deepen the formal understanding of such a concept. The pages of the book devoted to the emergence of life are in this sense particularly significant.

---

[83]See Part II, Note 9.
[84]Visetti (2004, pp. 240–245).
[85]Bickhard (2004, p. 82).
[86]Visetti (2004, p. 245).
[87]Idem, p. 231.
[88]See Stano and Luisi (2016, pp. 210–246).

3.   Emergence and Information

The emergence of life is one of the most discussed topics among biologists and physicists. Buccheri takes a position in this debate, not only introducing a clear criterion of distinction between *inert* and *living* matter, but also elaborating a specific hypothesis concerning the arising of life and its development.

Regarding the first step, the author employs a concept that has been very popular in many research fields. I refer to the concept of *information*. This notion has a long history in the domain of biological sciences,[89] and Buccheri makes of it the main distinctive feature between "living" and "not living":

> "A property, the *information*, emerging at a high level of complexity, [...] whose sudden appearance establishes the searched point of separation between (simple) inert matter and (complex) living matter."[90]

In other words, it is a firm conviction of the author that the emergence of life is strictly connected to the transmission of specific organizational constraints. The set of such regularities—or *information*—supports "the rigorous and coordinated organization of myriads of cells ($\sim 10^{27}$ atomi),"[91] and implies a goal that is "intrinsic to matter itself."[92] This last remark concerns the second step of Buccheri's argumentation.

If the information, or, more precisely, the informative property "like the formation of DNA"[93] should be inherent to matter,

> "We will not need any statistical calculations in order to show the probability of occurrence of this event."[94]

Hence, the formulation of a very audacious hypothesis: in so far as information is inherent to the matter, the biotic processes of life could emerge, in principle,

---

[89]See Atlan (1972), Yockey (1992), and Auletta (2011).
[90]See Section 2.6.
[91]Ibid.
[92]See Benz (2000, p. 138).
[93]See Section 2.6.
[94]Ibid.

"on every other planet in any other conditions with specific adaptation paths—obviously with possible different results in terms of physical configuration and dimensions concerning the living structures."[95]

So, this hypothesis is audacious in its challenge to others received but shows opposite points of view about the origins of life.

Presently, the explanation of such a phenomenon seems to be polarized between two radically alternative positions: (1) the life as the value of a *statistical accident* and (2) the life as the product of a creative act ex nihilo. It is known that these assumptions represent the horns of an epistemological dilemma on which any potential solution risks being impaled. According to the followers of each perspective—that is, *Darwinians* and *Creationists*—there is nothing in what the other says that can be justified from an empirical-theoretical view. Hence a vicious circle from which it is hard to escape. Well, the hypothesis formulated by Buccheri consents to mitigate the effects of such a cul de sac: by identifying the source of life with informational processes *internal* to the matter, the author can propose a third point of view that is different from both the Darwinian and the Creationist perspective. Put in more precise terms, he can introduce an *autopoietic* point of view according to which

"life would simply derive from the tendency of matter to self-organize until the formation of very complex molecules, finally leading to the appearance of the *informative* property."[96]

Briefly, the model proposed by Buccheri is the expression of a "third way" between two distinct visions of life that are unsatisfactory for different reasons: the Creationist perspective is untenable because it is outside of any kind of scientific normativity; the Darwinian one because it invokes in its explanations a not clearly defined notion of *randomness.*

Having said that, we can see another aspect of the autopoietic hypothesis that is necessary to gain knowledge. It concerns its *heuristic* potentialities. In defending the informational approach to

[95]Ibid.
[96]Ibid.

the emergence of life, Buccheri raises a question of great importance. He asks whether it is possible that the phenomenon of life may exist without carbon.[97] The answer is amazing. As it is asserted in the book,

"Even if our knowledge shows that the determinant element for life in the Earth is carbon, it is also true that silicon and boron are as well able to form long and complex molecular chains. It could therefore exist, at least in principle, forms of life based on these elements provided that they would find favorable physical-climatic conditions, although different from those on the Earth."[98]

Now, the heuristic potential of Buccheri's approach lies precisely in this "at least in principle." And this because such an expression suggests that there are new conceptual possibilities in developing unknown research domains, although

"of course, we are making hypotheses; any possibility of verification is out of our present knowledge and technology."[99]

But, as the author remarks,

"Nature is much richer and complex than us […] it could well exist—in the remote vastness of the universe or in the remaining four billions years of possible existence of our solar system—nature's achievements significantly exceeding both the fantasy and the intellectual abilities of a human being."[100]

In this sense, the conceptual path opened by Buccheri not only describes an emergentist point of view on the origin of life but also constitutes a terrain that embodies the *cognitive emergence* of new concepts and perspectives. Before going to the conclusions, and in accordance with this last remark, I would like to deepen the epistemological meaning of a concept employed by the author. I refer to the notion of emergence.

---

[97]See Section 2.3.
[98]Ibid.
[99]Ibid.
[100]Ibid.

4. Emergence and Reduction

In his approach, as I said, Buccheri substitutes a metaphysics of elementary particles for a metaphysics of processes. I find this step very useful and apt to accord an epistemological dignity to emergence. In the past 50 years, such a concept has been strongly criticized. Notably, many scholars inspired by reductionist programs argued that emergentism is untenable from every point of view.

Consider, for instance, the critics developed by Jaegwon Kim (1998, 1999). By attacking the concept of *downward causation*, the American philosopher seems to have greatly reduced the explanatory scope of emergence. The main steps of such a critical assessment are as follows:

1. Downward causation (henceforth referred to as DC), that is, "the concept that a system as whole has a causal influence on its constitutive parts,"[101] is one of the main features of emergence.
2. But, taken seriously, it seems a very paradoxical and redundant concept.
3. And this because DC is in conflict with two fundamental metaphysical principles: (1) "the causal power actuality principle"[102] (henceforth referred to as PAC) and (2) "the principle of explanatory exclusion"[103] (henceforth referred to as PEX).

Compared to PAC, DC raises a key question: "How is it possible for the whole to causally affect its constituent parts on which its very existence and nature depends?"[104]

With respect to PEX, DC raises the issue of explaining how "two causal stories about a single phenomenon mesh with each other."[105]

The search for appropriate answers to these theoretical problems would demonstrate, according to Kim, that DC "impl(ies) a kind of self-causation"[106] and

---

[101]Hulswitt (2006 p. 261).
[102]See Kim (1999).
[103]See Kim (1998).
[104]Kim (1999, p. 28).
[105]Kim (1998, p. 66).
[106]Kim (1999, p. 28).

"creates an instable situation requiring us to find an account of how the two purported causes are related to each other."[107]

So, for these reasons, DC would be a redundant and paradoxical concept.

Obviously, in Buccheri's perspective, such reasons are untenable and insufficient to justify a total rejection of the emergentist point of view. As it has been observed by many philosophers of biology and physics (e.g., Bich [2012]; Bitbol [2012]), the Kimian critics of DC suppose a metaphysics of elementary particles that has been completely disqualified by the research achieved in the fields of fundamental physics. Anyway, some aspects of such critics could be useful in order to better understand the nomic profile of emergent processes.

4.1   Bitbol as a Reader of Kim

It is not by chance that Bitbol (2012, pp. 234–235) accords great importance to Kim's thesis. This is the way in which he resumes the main argumentative steps of such a point of view on DC:

> "The central difficulty, expressed by Kim (1999), is the threat of vicious circularity. Is it coherent, Kim asks, to assume that the presence of a certain lower-level process is 'responsible' for the presence of a higher-level process, and yet the higher-level process somehow exercises a causal influence on the lower-level process? Since the higher level entirely arises from the lower level, the idea of downward causation sounds either contradictory or redundant. It contradicts the rest of science if it really makes a difference, by violating the micro-laws that rule the low-level. And it is redundant if it merely restates in a different language, a coarse-grained language, what could, at least in principle, be couched in the fine-grained language of the detailed micro-processes."

As one can see, this passage highlights both of the problems that DC has to face: The first one, that is, "self-causation," is explicated with particular reference to a common presupposition "of the scientific picture of the world." I refer to an epistemological

---

[107]Kim (1998, p. 65).

assumption known as "nomological closure of the lower micro-*physical* level"[108] or as "inclusivity of levels."[109] The second one, that is, "the presence of two causal stories," is, in turn, explicated with reference to another common epistemological presupposition known as "causal fundamentalism."

Briefly, putting aside technicalities, DC wouldn't fit a *naturalistic* description of the world, because it would seem to violate some fundamental constraints of its own "ontological furniture." First, by assuming that the *whole* (or the *macrolevel*) is able to alter its own *constituent parts* (or the *microlevel*), one introduces a model of "causal circularity" leading toward "an apparent absurdity."[110] Second, by supposing the co-existence of two causal paths, that is, the co-presence of an *upward* path and of a *downward* path, one could imply that a same phenomenon (or effect) "is causally overdetermined."[111] Therefore, Bitbol observes, if one want to maintain an emergentist view of nature, it is necessary to place DC in a new epistemological framework. Of course this suggestion is not new. Other philosophers of science have tried to recast DC by proposing innovative approaches to emergence. It is the case, for instance, of Emmeche et al. (2000): the distinction they have made between three versions of DC represents a great theoretical novelty. Moreover, their endeavor of describing the third one in terms of *stable structural attractors* has revived DC and the debate about emergence. It is, however, my firm conviction that it is to Bitbol that one owes the most original treatment of DC. His *relational* epistemology, as I shall try to demonstrate, is the ideal framework in which to examine DC and the model of co-emergence proposed by Buccheri.

5.   Buccheri and Bitbol: A Short Comparison

I will start by mentioning an interesting remark by Buccheri in Part III, "Premise":

---

[108]Bitbol (2012, p. 235).
[109]Emmeche et al. (2000 p. 18).
[110]Kim (1999, p. 28).
[111]Kim (1998, p. 65).

"If we consider that the classic physics was founded on the assumption to be an 'objective' description of reality, we may understand how dramatic was the crisis of the determinism and separation following the discovery of the quantum phenomena. A crisis that has invested also the concept of 'objectivity' when it is used to label the results of the scientific work, if we consider that the term 'objective' may not necessarily indicate something existing and acting independently from us."

I propose to construe what Buccheri says in this passage in terms of Bitbol (2012, pp. 653–654):

"Selon une épistémologie naturalisée des relations [. . .] il n'est pas question de supposer que les divers niveaux d'organisation [. . .], éléments de base ou caractéristiques globales émergentes, préexistent à la relation cognitive. C'est dans et par cette relation que des traits […] se *produisent*, et c'est donc chaque type de relation cognitive qui *définit* la nature et l'échelle des traits correspondants. […] Une telle inversion de l'ordre des priorités, de la primauté de l'objet à la primauté de l'acte ou de la relation épistémique, conduit bien à attribuer à un trait émergent un statut intermédiaire entre une véritable propriété intrinsèque et un vague épiphénomène: celui de corrélat d'un rapport cognitif ayant valeur constitutive."

This combination of theoretical perspectives is justified by two reasons. First, Bitbol's relational epistemology fits very well the general framework sketched by Buccheri: the "primacy of epistemic relation" (*la primauté* [. . .] *de la relation épistémique*) is what such a framework needs (and supposes) in order to criticize the classical physics approach to objectivity. Indeed, it is only by means of an "inversion of priorities" between "objects" and "relations" that one can assert that "the term 'objective' may not necessarily indicate something existing and acting independently from us."

Second, what Buccheri suggests, namely the dismissing of the classical concept of objectivity, is a theoretical move that every consistent emergentist approach implicitly demands. The relational epistemology of Bitbol is the instrument that permits to realize such a target.

5.1   Saving DC

The French scholar, known for his work in both the philosophy of physics and in the philosophy of the mind, aims at recasting the concept of emergence in the light of some aspects of the quantum revolution in natural sciences. Among these, three are worth considering: (1) the already mentioned priority of the "epistemic relation," (2) the substitution of the classical concept of "property" for the quantum one of "observable," and (3) the absence of a basic level composed of simple elements or individuals.

Such theoretical moves are an essential step in saving DC from the reductionist attack of Kim. The main critics addressed by the American philosopher base their arguments upon some of the mentioned concepts. For instance, the argument of the "nomological closure of the lower micro-*physical* level" works in so far as one supposes an ontology that reduces the microlevel to a set of interacting individuals, each of which is the bearer of "such and such" properties.

In this metaphysical framework, DC, or at least the strong version of it (henceforth referred to as SDC), doesn't fit the constraints of a naturalistic description. As I said, it implies a radical alteration of the microlevel, that is, a process that is *absurd* and *unrealistic*. It is absurd, because the macrolevel would seem to modify the microlevel *at the same time that* the microlevel realizes the macrolevel, which "looks like a bizarre metaphysical bootstrapping exercise."[112] It is unrealistic *because the macrolevel doesn't modify the microlevel*. On this last point, it is necessary remembering what Emmeche et al. (2000) say about the biology of cells.

5.2   A Case Study: The Cell's Biology

According to the three scholars, the relationships between the cell and its molecular constituents can be described neither in terms of *upward efficient causation* nor in terms of *downward efficient causation*. This double denial is justified by the "type" of causality at stake in SDC.

---

[112]Queiroz and Niño El-Hani (2006, p. 87).

In so far as it implies a temporal process characterized by the "*transfer* of some conserved quantity (say *energy*), from the cause to effect,"[113] efficient causation isn't a good model for understanding the cellular metabolism. This special kind of causal action fails to grasp the life of cells with respect to two main aspects: one concerning the presumed temporal order of events in the cell and the other concerning the biochemical bonds between molecules and cell. As for the first aspect, one could suppose that cells work as "a two-stage machine": the physicochemical reactions between the molecular constituents *realize* the biofunctional tissues of the cell (stage 1); the properties of the tissues so generated *retroact on* the "departure" properties of the molecules by converting the physicochemical reactions into biological processes as such (stage 2). *But this is not the case.* Contrary to what an explanation model inspired by efficient causality suggests, neither the physicochemical reactions between molecular constituents realize the biofunctional tissues of the cell, nor the properties of such tissues retroact on the "departure" properties of the molecules entailed in the biological process. So, this is enough to disqualify the vision of cells as "two-stage machines." Indeed, whatever is the meaning of verbs like "to realize" and "to retroact," it is clear that "A" *realizes* "B," if, and only if, "A" comes before "B" and that "B" *retroacts* on "A," if, and only if, "B" comes after "A." Nevertheless, neither one nor the other of two scenarios fit the biological reality of cells:

> "By considering a cell as an emergent entity on the biological level and its physical basis, this criticism of strong downward causation may be exemplified. In describing this emergent entity, we are very often tempted to use downward causation-like concepts in the strong sense; as if, for instance, the emergence of the cell as a living substance efficiently causes changes in the molecules, making them somehow specifically 'biological', i.e. substantially different from molecules of the non-living world, or, alternatively, as if the cell as such (efficiently) causes changes in the biochemical reactions among its constituents. But if we imagine a microscopic view of this alleged causal process, we will be unable to find any effective causality in the scenario."[114]

---

[113]Bitbol (2012, p. 235).
[114]Emmeche et al. (2000, pp. 20–21).

Therefore, one can think of cells as "synchronic machines" or, more correctly, as "biological machines," where "the upward and downward causes are not temporally distinct."[115]

Hence, the biochemical bonds between cells and molecular constituents have to be construed within an explanatory framework that is radically different from every model of causal explanation that is regulated by the "before-after" logic. This means that efficient causation, as a prototypical example of the "before-after" process, has to be removed from the explanations of cellular life. The remark according to which the macrolevel doesn't modify the microlevel (see Part 5.1) has to be understood in light of this theoretical move, not before having clarified the meaning of "to modify" and "to realizes" in the framework of SDC.

5.3   From SDC to WDC

If "to realize" means "to generate something," "to modify" refers to the action of converting something into something else. In the field of the cellular life, this last process would consist of transitions from some "departure" properties to some "destination" properties. Namely, it would consist of transitions from the physicochemical properties of certain molecular constituents to the biological ones of the same constituents in the framework of *organic life*.

But, as I said, this is not the case, because "the emergence of the cell as a living substance […] [doesn't] cause changes in the molecules, making them somehow specifically 'biological', i.e. substantially different from molecules of the non-living world."[116] Anyway, in the words of Michel Bitbol, "doit-on donner entiérement raison aux réductionnistes?"[117] *Absolutely not.* Even if SDC is disconfirmed by empirical evidences that are based on a cell's biology, there are other versions of DC fitting its organizational plan. Among these, the so-called "weak downward causation" (henceforth referred to as WDC) is one of "the viable candidates for a scientifically compatible account of DC."[118]

---

[115]Idem, p. 23.
[116]Ibid.
[117]Bitbol (2012, p. 656).
[118]Queiroz and Niño El-Hani (2006, p. 87).

5.3.1 *WDC: morpho-mereologies*

WDC possesses a great heuristic value that has been precisely proved in the domain of biology.[119] From a general point of view, it is a "case of formal synchronic causation."[120]

In other words, it is an ideal framework for grasping the *simultaneous* co-existence of different relational regimes. Naturally, such a kind of simultaneity corresponds to a class of processes that are the exact opposite of "two-stage machines." One can have a clear example of this difference by going back to examine the biochemical bonds that characterize cellular metabolism.

If in the perspective of SDC they can be grasped in terms of verbs like "to realize" and "to modify," in WDC they can be described in terms of predicates like "to be the part of" and "to be the pattern of." Now, the relations that respectively fall under the first and the second predicate are *temporally neutral* connections. And this is because neither one nor the other of such connections is subject to a "before-after" *logic*. For instance, the relation "to be the part of" is a temporally neutral connection because it is by no means possible to equate the assertion "A is the part of B" with the assertion "A comes before B." Of course the same goes for the second relation: the assertion "B is a pattern of A" cannot, in turn, be equated with the assertion "B comes after A." So, in so far as every cell *is constituted by* molecular constituents that are, in turn, *patterned* by the cell, one may conclude that the constituents are the *mereological* correlate of the cell and the cell the *morphological* correlate of its own constituents. The morphology of cell—its *relatedness*—is what constrains the behavior of the molecular constituents. In others words, it represents the "point of reference"—or the "attractor"—of its molecular microlevel. For the sake of simplicity, I cannot go further into this formal aspect of WDC, but I wish to remind you that concepts as those of "sinchronic relatedness" and "attractor" are sufficiently general to cover many research domains like the so-called phenomenon of "swarm intelligence"[121] or the biology of the brain.

---

[119]See Green (2018, pp. 998–1011).
[120]Queiroz and Niño El-Hani (2006, p. 87).
[121]See Kennedy et al. (2001).

Hence, contrary to what might be expected considering the critical view advanced by Kim on DC, emergence does fit the constraints of a naturalistic description by means of the constitution of synchronic morpho-mereologies regarding the structure of cells as well as one of other complex systems like social insects or neural dynamics of the brain.

5.4   Configurational Epistemologies

One could suppose that this is enough to clarify the epistemological role played by emergence in natural sciences. Bitbol's remarks mentioned before, however, lead me to pursue a short analysis of the main issues at stake.

First, attractive though it may be, WDC leaves open the question of exactly how to determine the modus operandi in which the macrolevel controls the microlevel. Second, assuming that this problem can be readily solved, it is necessary to evaluate DC by considering the *cognitive* processes involved "dans l'acte de mise en relation des phénomènes."[122] The two points are strictly related. When I say that it is not clear how the macrolevel controls the microlevel, I refer to an epistemological problem that is possible to formulate as follows: in WDC's perspective, the macrolevel is supposed to act as a *global* whole, of which, nevertheless, the impact on the microlevel is not adequately explained. This explanatory difficulty is a consequence of a metaphysical presupposition that WDC—as well as every DC's variant[123]—shares with the reductionist program of Kim. This presupposition concerns the *differences* between macrolevel and microlevel: in WDC's framework, the whole is a *form*, that is, "un réseau bien défini de relations entre des constituants de base,"[124] which are, in turn, conceived as *individuals* or *monadic elements.* So, the problem is to understand how a *form* can act on *monadic elements.* Certainly, the conceptual repertoire of dynamical systems theory frames the high-level constraints imposed by a global form on its low-level constituents in terms of attractors and basins of attraction. But this kind of formal description is not enough to answer the question of how the macrolevel *directly acts* on

---

[122]Bitbol (2012, p. 661).
[123]Or at least those DC's variants managing the classical concepts of objectivity.
[124]Idem, p. 658.

the microlevel. You can specify the low-level behavior of elementary constituents in terms of *orbits or trajectories* in a phase-space, and you can specify the high-level constraints of the macrolevels in terms of attractors that determine the motion of such trajectories. It is a modeling strategy that pervades many emergentist epistemologies. However, although it is *descriptively* powerful, this strategy is unable to *explicate* how "a configuration may act *as such* on a layer of basic properties."[125]

Given the differences between microlevel and macrolevel,

"It is still dubious whether stating [. . .] the global attractor and self-organizing constituents, is more than a verbal (or heuristic) device."[126]

The conceptual moves proposed by Bitbol (see Part 5.1) find their justification in the necessity of giving a realistic explanatory framework for the relationships between macrolevel and microlevel. Going backward, it is first and foremost the last move that allows us to develop a plausible account of such relationships: by denying the existence of basic elements or, more correctly, by acknowledging that

"chaque candidat actuel au titre de 'base' n'apparaît pas moins *configurationnel* que le niveaux d'organisation qu'il 'sous-tend',"[127] "it is possible to reset 'toute asymétrie entre les composants et le comportements émergents'."[128]

One can appreciate this approach by envisaging the theoretical risks entailed by a misknowledge of it. It is the case of Terrence Deacon. The American scholar is well known for the distinction he makes between high-level emergent forms (or *topological configurations*) and low-level monadic elements (or *basic constituents*). Nevertheless, in so far as he doesn't conceive the basic constituents in terms of *configurational* entities, he sharply separates one level from the other. Thus, as an immediate consequence of such a clear-cut demarcation, the macrolevel and microlevel are completely independent. This reciprocal independence leads to a potential dualism in which

---

[125]Bitbol (2012).
[126]Ibid.
[127]Bitbol (2012, pp. 660–661).
[128]Idem, p. 659.

"les constituants de base se voient assigner une position à part, puisqu'ils restent à l'abri de toute action exercée par l'organisation de niveau supérieur."[129]

Of course, WDC is safe from such criticism: the mereological connections that it allows mitigate the potential dualism underlying the differences between microlevel and macrolevel. In any case, for the reasons set out, it is clear that a configurational construal of low-level constituents is also what WDC needs to achieve a consistent emergentist description of the world.

But introducing a configurational approach to basic elements would seem to imply the elimination of any difference between microlevel and macrolevel. This is precisely the case in Bitbol's perspective. He boldly asserts that

"il n'y a aucune différence essentielle entre le niveau qu'on suppose 'fondamental' et les niveaux émergents."[130]

The author supports this assertion by making appeal to some important results in *quantum fields theory*. First, he observes that

"la théorie quantique des champs [. . .] ne concerne en effet d'aucune manière des individus localisés traités comme 'substances', mais seulement des *types de configuration*."[131]

Second, he generalizes this remark by claiming that

"tous les niveaux d'organisation tombant dans le domaine de la physique se manifestent comme autant de configurations."[132]

I am not able to assess the appropriateness of this epistemological strategy. In particular, I am not sure that it is correct to extend a local frame to the whole of theoretical physics. Nevertheless, by assuming the feasibility of such an operation, another question has to be discussed, namely whether it is still possible to talk about "emergence" in a generalized configurational framework.

I raise the question because of the usual way of defining "emergence." Generally, it is presented as

---

[129]Idem, p. 658.
[130]Idem, p. 659.
[131]Ibid.
[132]Ibid.

"an outcome of local interactions between a large number of components [. . .] lead(ing) to the formation of spatio-temporal patterns [. . .] at global or macro-level."[133]

Of course, it is not the unique way, but it is undoubtedly one of the most common. Indeed, contrary to what Bitbol suggests, it presupposes a clear-cut distinction between microlevel and macrolevel. Hence, if framed within the boundaries of the configurational approach, a concept so defined risks to verse in pure nonsense. Fortunately, it isn't so.

Despite the collapse of the difference between microlevel and macrolevel, it is still possible to talk about "emergence." But this depends on how one interprets "collapse." There is a sense of the word that seems to divest emergence of any nomological significance. I refer to "collapse" understood as synonymic of "total reset." Yet this is not the sense in which Bitbol would mean the word. In his configurational approach, he aims not so much at annihilating the difference between microlevel and macrolevel as at dynamically modulating the boundaries between them. I say "dynamically modulating" to mean that for Bitbol such boundaries are not fixed once and for all. Indeed, they change, and such a fact fits very well the nomological significance of emergence.

Naturally, I refer to an idea of emergence in which *epistemic relations* and *quantum observables* play a pivotal role. These concepts are central in Bitbol's perspective because they allow to account for the *dynamicity* of relationships between microlevel and macrolevel. Simply put, they clarify that "what counts as a whole" and "what counts as a part" are always a result of complex negotiations between many interacting factors. Among these, the *mathematical praxis* has a double "structuring function." That is to say, it is "interfacial"[134] and "transcendental."

"Interfacial," in so far as it is a source of symbolic representations in which embodied cognitive abilities, collective formal heritages, and experimental devices interact with a heterogeneous repertoire of phenomena. "Transcendental," in so far as such interactions are ""constitutive" (in the Kantian sense) of any and all objectivity." Well, it is this interactive network that molds emergence and that implies, on the one hand, an *interfacial* activity of modeling and, on the other,

---

[133]Lee (2012, p. 126).
[134]See Bitbol (2012, pp. 455–475).

a *transcendental* activity aiming at *defining* sets of "phénomènes expérimentaux possibles"[135] and at *attaching* meanings to some aspects of such phenomena "*sous conditions* d'utilisation d'une certaine classe de dispositifs expérimentaux."[136]

Therefore, far from being defined as "noyau d'*être* irréductible aux constituants,"[137] emergence is viewed by Bitbol as the "corrélat d'un rapport cognitif ayant valeur constituive"[138] and entailing an observative gap (or "résidu") between "what counts as a whole" and "what counts as a part" within the boundaries of historically determined interfacial activities. DC is the name of such a gap, or, more precisely, the name of an interfacial activity in which "a high-level coarse-grained device […] acts selectively at the [macro-level], in order to *modify* what is observed at the other level."[139] In conclusion, in this new framework, DC, or, more precisely, any DC's variant (weak, strong, etc.) is no longer understood as the *direct* action of the macrolevel on the microlevel but, rather, as a particular *observer activity.* Namely, as high-level formal specifications that allow to appreciate the differences between the *behavior of isolated material parts* and the *behavior of integrated functional components.*[140]

6. Conclusions: Dialectics of Emergence

I held that Bitbol's approach to emergence is a good framework within which to construe the dismissing of classical objectivity. I don't know whether Buccheri would adhere to the conceptual move of the French scholar. Nevertheless, it is my firm belief that he could be attracted by some aspects of Bitbol's configurational turn.

Surely, as I said, he could be persuaded to adopt a relationally inspired point of view. You can find significant traces of such a relational attitude in several thematic loci of *Myth, Chaos, and Certainty: Notes on Cosmos, Life, and Knowledge.*

But, among these, there is a passage that sounds particularly interesting. Let us read it: "A 'complex' thought where the concepts

---

[135]Idem, p. 648.
[136]Ibid.
[137]Idem, p. 667.
[138]Idem, p. 654.
[139]Bitbol (2007, p. 320).
[140]See Bich (2012, p. 336).

of unity and separation may not be reciprocally incompatible if the interpretation of every single fact or event would take the charge to always maintain a strict connection with the general context and its parts, as it should happen in the social case, for any specializations of the knowledge, born from an aware abstraction from their context with the aim to deepen the details."[141]

Although it doesn't appeal to either of the concepts employed by Bitbol, it correctly exemplifies how the *interfacial* and *transcendental* activities underlying the process of human knowledge operate in every endeavor of managing the multiform complexity of the world.

I refer to what the author says about "an aware abstraction from their context at the aim to deepen the details." This gnoseological operation supposes both a modeling strategy and a constitution process "in the Kantian—and Bitbolian—sense of the term." In short, it is interfacial and transcendental. What is interesting is that it constitutes an instance of what the French philosopher would call a "micro-analyse qui conditionne la réduction."[142] In fact, in pointing out the necessity of exploring the instances of "complexity thinking," Buccheri pays attention to what you can envisage as an epistemological attitude that is the exact opposite of "complexity thinking," namely "reductionist thinking," which aims at describing the local behavior of single elements.

If within the boundaries of classical objectivity this conceptual move is a highly contradictory step, in Bitbol's perspective it represents a *transcendental* (*and physiological*) *phase* of an interfacial activity. This point needs a little bit of explanation. If you define emergence by abstracting from the *cognitive-embodied-historic import* of mathematical praxis, it is clear that "an aware abstraction of an element from its context" is inconsistent with the previously mentioned "complexity thinking." In a realistic (or *mind-independent*) framework so defined, "what counts as a whole" is made of systemic constraints that are thought as global properties irreducible to local properties *such-and-such*. Therefore, the theoretical gesture that temporarily aims at envisaging the local behavior of the constituents is considered an act annihilating the systemic solidarities (or *cohesiveness*) of the whole. In this sense, it can't but represent a source of "systemic inconsinstencies" that

---

[141]See Section 6.1.
[142]Bitbol (2012, p. 667).

destabilize the nomic profile of emergence. But, if you change framework, this is not necessarily the case. For instance, if you inscribe emergence in a complex network of modeling activities, you can save the cohesiveness of the whole by at the same time assuming the epistemological relevance of theoretical gestures as the reductionist one.

It is possible to understand this conceptual move by going back to Bitbol's lesson. In his transcendental framework—that I would like to call "praxeological"—emergence is not defined in terms of irreducible properties *such-and-such*. Rather, it is presented as a descriptive gap between "what counts as a whole" and "what counts as a part" with respect to different "acte(s) de définition de complexes constants de relations entre des phénomènes."[143]

These acts have "valeur constitutive" and presuppose each other: the act that constitutes the "whole" presupposes the act that constitutes the "part," and vice versa. In this sense, "part" and "whole" are relational aspects of the same phenomenon (or of different phenomema) that "co-adviennent"—in Visetti's words—within formal praxeologies defining "what counts as a whole" with respect to "what counts as a part," and vice versa.

Hence, in a praxeological perspective, the reductionist gesture discussed by Buccheri doesn't destabilize the nomic profile of emergence. If you define it in terms of the mentioned descriptive gaps, you need to know, among other things, "what counts as a part" with respect to "what counts as a whole." And what guarantees this local knowledge is precisely the reductionist gesture. So, far from destabilizing the nomic profile of emergence, this gesture *dialectically* completes it. Indeed, "l'émergentisme ne peut pas se passer du réductionnisme comme *partenaire dialectique*."[144] This is, I believe, the great epistemological lesson that the reader will deduce from *Myth, Chaos, and Certainty: Notes on Cosmos, Life, and Knowledge*. And this will be, as I hope, the point of departure of Buccheri's forthcoming books.

Francesco La Mantia

Department of Humanistic Sciences
University of Palermo
Palermo, Italy

---

[143]Idem, p. 661.
[144]Idem, p. 667.

Appendix

L-systems are systems of parallel algorithms computing the biological development of simple filamentous organisms. For instance, let's consider a computational system (or "evolutive Lyndenmayer grammar") constituted by the following data:

1. An *alphabet* of the form: **Σ=<F,+,–, [], A>;**
2. A *rewriting rule* of the form: p_1**: A→F [+A] [–A] FA**.

So, under a given interpretation of such symbols, two simple iterations of p_1 generate a *string of words*, or a *computational pattern*, standing for a *composite geometric structure* that "capture" the first stages of development of a simple filamentous organism. Put in more concrete terms, if we decide that:

(1) **F** means "take a step forward in the direction in which you are looking, and leave a *graphic trace*";

(2) **+** means "rotate clockwise by 90°;

(3) **–** means "rotate anticlockwise by 90°;

(4) **[]** stands for "eventual ramifications" of the vegetal organism; and

(5) **A** stands for "meristematic apex," that is, for "a segment of vegetal tissue generating other segment of the same organism."

The iterations of p_1, visualized on a *graphic display*, generate the following geometric *leaf* (see the figure here).

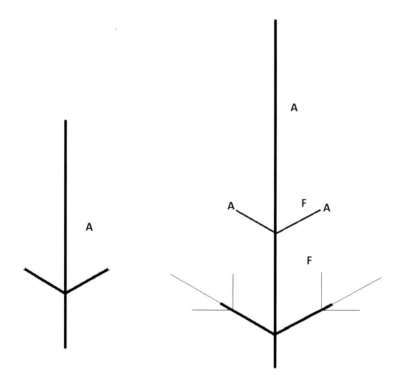

Naturally, *iterative processes* (see Kauffman [2015, p. 55]; Hägeräll et al. [2015, p. 81]) play a fundamental role in the *patterned form* of computational and biological structures.

Bibliography

Abbott, E. A. *Flatlandia*, Adelphi, Milano, 1998 (*Flatland*, 1882).

Alfano, M., Buccheri, R. Il Symbolon fra le Simplegadi di Mito e Logos, in *Il Simbolo nel Mito attraverso gli studi del Novecento*, Recanati-Ancona, 2006.

Alfano, M., Buccheri, R. Il Symbolon come oltrepassamento nel Logos di Mito e Scienza, in *Il Mito in Sicilia*, a cura di G. Romagnoli, Palermo, 2007.

Alfano, M., Buccheri, R. *L'energia creativa dell'oscillazione polare fra mito e scienza*, Saladino, Palermo, 2009.

Alfano, M., Buccheri, R. *Oltre la razionalità scientifica*, Lateranum, LXVIII, Roma 2012.

Aristoteles. *Metaphysics*, A2, 982b, 12–20.

Ascher, M. *Etnomatematica: Esplorare concetti in culture diverse*, Boringhieri, Torino, 2007 (*Mathematics Elsewhere: An Exploration of Ideas Across Cultures*, Princeton University Press, 2002).

Atlan, H. *L'organisation biologique et la théorie de l'information*, Seuil, Paris, 1972.

Auletta, G. *Cognitive Biology: Dealing with Information from Bacteria to Minds*, Oxford University Press, Oxford, 2011.

Aveni, A. *Conversando con i pianeti: Il Cosmo nel mito e nella scienza*, Dedalo, Bari, 1994 (*Conversing with the Planets: How Science and Myth Invented the Cosmos*, 1992).

Ayala, F. J. *Le ragioni dell'evoluzione*, Di Renzo Editore, Roma, 2005.

Barrow, J. *I numeri dell'universo: Le costanti di Natura e la Teoria del Tutto*, Mondadori Printing SPA, Cles (TN), 2004 (*The Constants of Nature: From Alpha to Omega*, Jonathan Cape, 2002).

Barrow, J., Tipler, F. *The Anthropic Cosmological Principle*, Oxford University Press, Oxford, 1986.

Bènard, H. Les tourbillons cellulaires dans une nappe liquide transportant de la chaleur par convection en règime permanent, *Ann. Chim. Phys.*, **7**(23), 62–144, 1901.

Benz, A. *The Future of the Universe: Chance, Chaos, God*, Continuum, New York, London, 2000.

Bich, L. *L'ordine invisibile: Organizzazione, autonomia e complessità del vivente*, Rubbettino, Soveria Mannelli, 2012.

Bickhard, M. The dynamic emergence of representation, in *Representation in Mind: New Approaches to Mental Representation*, eds., Clapin, H., Staines, P., Slezak, P., Elsevier Science, 2004.

Bitbol, M. Ontology, matter and emergence, *Phenom. Cogn. Sci.*, **6**, 293–307, 2007.

Bitbol, M. *De l'intérieur du monde. Pour une philosophie et une épistémologie des relations*, Flammarion, Paris, 2012.

Bocchi, G., Ceruti, M. *Origini di storie*, Feltrinelli, Milano, 2006.

Boella, L. *Empatie: L'esperienza empatica nella società del conflitto*, Raffaello Cortina, Milano, 2018.

Böhm, D. *Wholeness and the Implicate Order*, Routlege, London, New York, 2008.

Böhm, D., Peat, D. *Science, Order, and Creativity*, Routledge, London, New York, 2005.

Boltzmann, L. On certain questions of the theory of gases, *Nature*, **51**, 413–415, 1895.

Boniolo, G. *Il limite e il ribelle*, Raffaello Cortina, Milano, 2003.

Boniolo, G. *Il pulpito e la piazza: Democrazia, deliberazione e scienze della vita*, Raffaello Cortina, Milano, 2011.

Boriakoff, V., Buccheri, R., Fauci, F. Discovery of a 6.1 millisecond binary pulsar, PSR1953+29, *Nature*, **304**, 417–419, 1983.

Bria, P., Oneroso, F. *La bilogica fra mito e letteratura: Saggi sul pensiero di Matte Blanco*, Franco Angeli, Milano, 2015.

Buccheri, R. Time and the dychotomy subjective/objective. An endophysical point of view, Russian Temporology Seminar, Moscow, 2002, http://www.chronos.msu.ru/EREPORTS/

Bucheri _time.htm.

Buccheri, R. Temporalità e interazione. Dall'apofantico …, in *Tempo della fisica e Tempo dell'uomo: Relatività e relazionalità*, Akousmata • orizzonti dell'ascolto, Trapani-Ferrara, 2008.

Buccheri, R. La dualità dell'uomo tra fede e scienza. Dalla neurofisiologia alla letteratura, *Dialoghi Mediterranei*, n.19, 2016.

Buccheri, R. L'intreccio pirandelliano di istinto e ragione, *Dialoghi Mediterranei*, n.26, 2017.

Buccheri, M., Buccheri, R. Evolution of human knowledge and the endophysical perspective, in *Endophysics*, *Quantum and the Subjective*, eds., Buccheri, R., Elitzur, A. C., Saniga, M., World Scientific Publishing, Singapore, 2005a.

Buccheri, M., Buccheri, R. Le antiche cosmogonie e la moderna cosmologia: Il mito come forma primaria di conoscenza, in *Aspetti e forme del Mito: la sacralità*, Edizioni Anteprima, Palermo, 2005b.

Campbell, D. T. The philosophy of Karl Popper, in *Evolutionary Epistemology*, Open Court, La Salle, IL, 1974.

Campbell, J. *Mito e modernità: Figure emblematiche di un passato antichissimo nell'esperienza quotidiana*, Edizioni red!, Milano 2007 (*The Impact of Science on Myth*, 1961).

Capra, F. *Il Tao della Fisica*, Adelphi, Milano, 1993 (*The Tao of Physics: An Exploration of the Parallels between Modern Physics and Eastern Mysticism*, Shambala, Berkeley, 1975).

Capra, F. *La rete della vita: Una nuova visione della natura e della scienza*, RCS Libri, Milano, 1997 (*The Web of Life: A New Scientific Understanding of Living Systems*, Doubleday, New York, 1996).

Carter, B. *IAU Symposium*, n.63, 1974.

Craik, K. *The Nature of Explanation*, Cambridge University Press, 1943.

Cusumano, A. *Antropologia delle migrazioni*, Istituto Euroarabo, Mazara del Vallo, 2016.

Cusumano, A. *Monoteismi e dialogo*, Istituto Euroarabo, Mazara del Vallo, 2017.

Damasio, A. *Emozione e coscienza*, Adelphi, Milano, 2003.

Damasio, A. *L'errore di Cartesio. Emozione, ragione e cervello umano*, Adelphi, Milano, 2008 (*Descartes' Error: Emotion, Reason, and the Human Brain*, Putnam Publishing, 1994).

Darwin, C. *L'origine delle specie: Selezione naturale e lotta per l'esistenza*, Boringhieri, Torino, 1967.

Deane-Drummond, C., Szerszynski, B., Grove-White, R. Introduction to Part II, in *Reordering Nature: Theology*, *Society and the New Genetics*, T & T Clark, Continuum Imprint, London, NewYork, 2003.

de Duve, C. *Polvere vitale: Origine ed evoluzione della vita sulla Terra*, Mondolibri, Milano, 1998 (*Vital Dust: The Origin and Evolution of*

Life on Earth, Basic Books, a division of Harper Collins Publisher, Inc., 1995).

De Felice, F. *Cosmogon: Alla ricerca delle ragioni dell'essere*, Akousmata • orizzonti dell'ascolto, Ferrara, 2012.

Deutsch, D. *La trama della realtà*, Biblioteca Einaudi, 1997 (*The Fabric of Reality*, Viking Press, 1997).

de Waal, F. *Il bonobo e l'ateo*, Raffaello Cortina, Milano, 2017 (*The Bonobo and the Atheist: In Search of Humanism Among the Primates*, W. W. Norton & Company, 2013).

Dürrenmatt, F. *La promessa: Un requiem per il romanzo giallo*, Feltrinelli, Milano, 1959 (*Das versprechen: Requiem auf den Kriminalroman*, Diogenes Verlag AG, Zurich, 1985).

Dürrenmatt, F. *Nel cuore del pianeta*, Marcos Y Marcos, Milano, 2003 (*Nachgedanken*, Diogenes Verlach AG, Zürich, 1998).

Eigen, M. *Gradini verso la vita: L'evoluzione prebiotica alla luce della biologia molecolare*, Adelphi, 1992 (*Stufen zum leben: Die frühe Evolution im Visier der Molekularbiologie*, R. Piper GHBA & Co. KG, München, 1987).

Einstein, A., Born, M. *Scienza e vita: Lettere 1916–1955*, Einaudi, Torino, 1973.

Ellenjimittam, A. *La quintessenza delle religioni*, Verdechiaro, Baiso (RE), 2001.

Emmeche, C., Køppe, S., Stjernfelt, F. Levels, emergence and three versions of downward causation, in *Downward Causation: Mind, Bodies and Matter*, eds., Andersen, P. B., Emmeche, C., Finneman, N., Voetmann, P., Aahrus University Press, Aahrus, 2000.

Figà-Talamanca Dore, L. *La logica dell'inconscio: Introduzione all'opera di Ignacio Matte Blanco*, Edizioni Studium, Roma, 1978.

Filoramo, G., et al. *Manuale di storia delle religioni*, Mondolibri, Milano, 1998.

Fischer, E. P. *Aristotele, Einstein e gli altri*, Raffaello Cortina, Milano, 1997 (*Aristoteles, Einstein & Co.*, R. Piper, GmbH & Co. KG, München, 1995).

Frova, A. *Armonia celeste e dodecafonia*, BUR Rizzoli, Milano, 2006.

Gadamer, H. G. *Verità e metodo*, Bompiani, Milano, 1983.

Gleiser, M. *Il Neo del Creatore: L'imperfezione nascosta nel miracolo della vita*, Rizzoli, Milano, 2011 (*A Tear at the Edge of Creation: A Radical New Vision for Life in an Imperfect Universe*, Free Press, 2010).

Goethe, J. W. *Le affinità elettive*, Mondadori, Milano, 1988 (*Die Wahlwewandtshaften*).

Goldberg, E. *Il paradiso della saggezza: Come la mente diventa più forte quando il cervello invecchia*, Ponte alla grazie, Milano, 2005 (*The Wisdom Paradox: How Your Mind Can Grow Stronger as Your Brain Grows Older*, Avery, 2006).

Gomarasca, P. *La ragione negli affetti: Radice comune di Logos e Pathos*, Vita e pensiero, Milano, 2007.

Gould, S. J. *Punctuated Equilibrium*, Belknap Press, 2007.

Green, S. Scale-dependency and downward causation in biology, *Philos. Sci.*, **85**(5), 998–1011, 2018.

Hack, M., Battaglia, P., Buccheri, R. *L'idea del Tempo*, UTET, Torino, 2005.

Hägeräll, C., Taylor, R., Cervén, G., Watts, G., Van de Bosch, M., Press, D., Minta, S. Biological mechanisms and neurophysiological responses to sensory impact from nature, in *Oxford Textbook of Nature and Public Health*, ed., Van De Bosch, M., Minta, S., Oxford University Press, Oxford, 2015.

Hawking, S. *Dal Big Bang ai Buchi Neri: Breve storia del Tempo*, Rizzoli, Milano, 1997 (*A Brief History of Time: From the Big Bang to Black Holes*, Bantam Dell Publishing Group, 1988).

Hawking, S. *Buchi neri e universi neonati: Riflessioni sull'origine e sul futuro del cosmo*, Rizzoli, Milano, 1993.

Heidegger, M. *Nietzsche*, ed., Volpi, F., Biblioteca filosofica, Adelphi, Milano, 2000.

Heller, M. *Tensione Creativa: Saggi sulla scienza e sulla religione*, Akousmata • orizzonti dell'ascolto, Ferrara, 2012 (*Creative Tension: Essays on Science and Religion*, Templeton Foundation Press, Pennsylvenia, 2003).

Hesse, H. *Narciso e Boccadoro*, Mondadori, Milano, 1989 (*Narziss und Goldmund*, Medusa, 1933).

Hillman, J. *La vana fuga dagli dei*, Adelphi, Milano, 2005 (*On Paranoia & On the Necessity of Abnormal Psychology: Ananke and Athena*, Eranos Jahrbuch, LIV, 1985 & XLIII, 1974).

Hulswitt, M. How causal is downward causation?, *J. Gen. Philos. Sci.*, **36**, 261–287, 2006.

James, W. *The Varieties of the Religious Experience*, Longmans, London, 1902.

Jastrow, R. *God and the Astronomers*, Warner Books, New York, 1978.

Jauch, J. M. *Sulla realtà dei quanti: Un dialogo galileiano*, Adelphi, Milano, 2001 (*Are Quanta Real?: A Galileian Dialogue*, Indiana University Press, 1973).

Johnson-Laird, P. *Modelli Mentali*, Il Mulino, Bologna, 1988 (*Mental Models: Toward a Cognitive Science of Language, Inference and Consciousness*, Cambridge University Press, 1983).

Jung, C. G. *Il problema dell'inconscio nella psicologia moderna*, Einaudi, Torino, 1973.

Jung, C. G., Kerényi, K. *Prolegomeni allo studio scientifico della mitologia*, Boringhieri, Torino, 2012 (*Einführung in das Wesen der Mythologie*, Pantheon Akademische Verlangsanstalt, Amsterdam-Leipzig, 1942).

Kauffman, S. *Unified Reality Theory*: *The Evolution of Existence into Experience*, Balboa University Press, 2015.

Kennedy, J., Russell, E., Yuhui, S. *Swarm Intelligence*, Morgan Kauffman Publishers, San Francisco, 2001.

Kim, J. *Mind in a Physical World: An Essay on the Mind-Body Problem and Mental Causation*, MIT Press, Cambridge Massachussets, 1998.

Kim, J. Making sense of emergence, *Philos. Stud.*, **95**, 3–36, 1999.

Lee, C. Bounded rationality and the emergence of simplicity admist complexity, in *Nonlinearity*, *Complexity and Randomness in Economics: Towards Algorithmic Foundations for Economics*, eds., Zambelli, S., George, D. A. R., Wiley-Blackwell, Oxford, 2012.

Leopardi, G. *Zibaldone di pensieri*, Garzanti, Milano, 1991.

Lévy-Bruhl, L. *La mitologia primitiva*, Newton Compton, Roma, 1973 (*La mythologie primitive: Le monde mythique des australiens et des papous*, Alcan, Paris, 1973).

Lévi-Strauss, C. *Mito e significato: L'antropologia in cinque lezioni*, Il Saggiatore, Milano, 2002.

Lévi-Strauss, C. *L'antropologia di fronte ai problemi del mondo moderno*, Bompiani, Milano, 2017 (*L'Anthropologie face aux problèmes du monde moderne*, Editions du Seuil, 2011).

Livio, M. *La sezione aurea: Storia di un numero e di un mistero che dura da tremila anni*, Rizzoli, Milano, 2012.

Lovelock, J. *Gaia: A New Look at Life on Earth*, Oxford University Press, Oxford, 1979.

Lyndenmayer, A. Automata, formal languages and developmental systems, in *Logic, Methodology and Philosophy of Science*, eds., Suppes, P., Henkin, L., Joja, A., et al., North-Holland, Amsterdam, 1971.

Maimònide, M. *La guida dei Perplessi*, Parte III, Cap. LI, Pos. 14399.

Martins, Z., et al. *Indigenous amino acids in primitive CR meteorites*, 2008 (arxiv.org/abs/ 0803.0743).

Marx, K., Engels, F. *La concezione materialistica della storia*, Editori Riuniti, Le idee, 1966.

Matte Blanco, I. *L'inconscio come insiemi infiniti: Saggio sulla bilogica*, Einaudi, Torino, 2000, trad it. di Pietro Bria (*The Unconscious as Infinite Sets: An Essay in Bilogic*, Gerald Duckworth & Co. Ltd., London, 1975).

Maturana, H. R., Varela, F. *Autopoiesi e cognizione: La realizzazione del vivente*, Marsilio, Venezia, 1992 (*Autopoiesis and Cognition: The Realization of the Living*, Reidel Publishing Co., Dordrecht, Holland, 1980).

Maturana, H. R., Varela, F. *Autocoscienza e realtà*, Raffaello Cortina, Milano, 1993 (*The Biological Foundation of Self-Consciousness and the Physical Domain of Existence*).

Monod, J. *Il caso e la necessità: Saggio sulla filosofia naturale della biologia contemporanea*, Mondadori, Milano, 1986 (*Le hazard et la nécessité*, 1970).

Montefoschi, G., Nierenstein, F. *Un solo Dio, tre verità. Arabi, Ebrei e Cristiani: l'enigma della fede*, Mondadori, Milano, 2001.

Nagel, E., Newman, J. R. *La prova di Gödel*, Boringhieri, Milano, 2000 (*Gödel's Proof*, New York University Press, New York, 1968).

Ong, W. J. *Oralità e scrittura: Le tecnologie della parola*, Il Mulino, Bologna, 1986 (*Orality and Literacy: The Technologizing of the Word*, Routledge & Kegan, London 1982).

Ortoli, S., Witkowski, N. *La vasca di Archimede: Piccola mitologia della scienza*, Mondolibri, Milano, 1998 (*La baignoire d'Archimède*, Èditions du Seuil, 1996).

Osnato, A. *L'utopia ecumenica di Padre Anthony Ellenjimittam*, Zeta Printing, Palermo, 2016.

Osserman, R. *Poesia dell'universo: L'esplorazione matematica del cosmo*, Longanesi, Milano, 1997 (*Poetry of the Universe: A Mathematical Exploration of the Cosmos*, Anchor Books, New York, 1995).

Penrose, R. *Il grande, il piccolo e la mente umana*, Raffaello Cortina, Milano, 2000 (*The Large, the Small and the Human Mind*, Cambridge University Press, 1999).

Peretz, I., Zatorre, R. J. *The Cognitive Neuroscience of Music*, Oxford University Press, Oxford, 2003.

Petitot, J. Centrato/acentrato, in *Enciclopedia Einaudi*, Vol. 3, Einaudi, Torino, 1979.

Pirandello, L. *L'umorismo*, Garzanti, Milano, 1995 (Foreword by Pietro Milone).

Pizzorno, A. *Sulla maschera*, Il Mulino, Bologna, 2008.

Platone, *Fedro*, a cura di Giovanni Reale, CDE, Milano, 1998.

Planck, M. *La conoscenza del mondo fisico*, Torino, Boringhieri, 1964.

Popper, K. R. *Congetture e confutazioni*, Il Mulino, Bologna, 1972.

Popper, K. R. *Il mondo di Parmenide: Alla scoperta della filosofia presocratica*, Mondolibri, Milano, 2001 (*The World of Parmenides*, Routledge, London, 1998).

Popper, K. R., Lorentz, K. *Il futuro è aperto*, Rusconi, Milano, 1989.

Prigogine, I., Stengers, I. *La nuova alleanza. Metamorfosi della scienza*, Einaudi, Torino, 1999 (*La nouvelle alliance. Métamorfose de la science*, Gallimard, Paris, 1979).

Prigogine, I. *La fine delle certezze: Il Tempo, il Caos e le leggi della Natura*, Boringhieri, Torino, 1997 (*La fin des certitudes: Temps, chaos et les lois de la nature*, Editions Odile Jacob, Paris, 1996).

Prusinkiewicz, P. Visual models in morphogenesis, in *Artificial Life*, ed., Langton, C., MIT Press, Massachussets, 1995.

Queiroz, J., Niño El-Hani, C. Semiosis as an emergent process, *Transactions of The Charles Sanders Peirce Society*, **42**(1), 78–116, 2006.

Remotti, F. *Prima lezione di Antropologia*, Laterza, Roma-Bari, 2017.

Rivera, N. *The Earth is Our Home: Mary Midgley's Critique and Reconstruction of Evolution and its Meanings*, Imprint Academic, Exter, 2010.

Rizzolatti, G. Corrado Sinigaglia, *So quel che fai: Il cervello che agisce e i neuroni-specchio*, Raffaello Cortina, Milano, 2006.

Rössler, O. E. *Endophysics: The World as an Interface*, World Scientific Publishing, London, 1998.

Russo, L. *La rivoluzione dimenticata: Il pensiero scientifico greco e la scienza moderna*, Feltrinelli, Milano, 2008.

Saniga, M. Pencils of conics: a means toward a deeper understanding of the arrow of time?, *Chaos, Solitons Fractals*, **9**, 1071–1086, 1998.

Schnitzler, A. *Doppio Sogno*, Adelphi, Milano, 1977 (*Traumnovelle*, S. Fisher Verlag, AG Berlin, 1931).

Schröder, G. *L'Universo sapiente: Dall'atomo a Dio*, Il Saggiatore, Milano, 2002 (*The Hidden Face of God*).

Schrödinger, E. *Che cos'è la vita?*, Adelphi, Milano, 1995 (*What is Life? The Physical Aspect of the Living Cell*, Cambridge University Press, 1944).

Smolin, L. *La vita del Cosmo*, Mondolibri, Milano, 1998 (*The Life of the Cosmos*, Oxford University Press, New York, Oxford, 1997).

Stano, P., Luisi, P. L. Theory and construction of semi-synthetic minimal cells, in *Synthetic Biology Handbook*, ed., Nesbeth, D. N., CRC Press, 2016.

Tanzella-Nitti, G. 2002, http://disf.org/principio-antropico.

Tattersall, I. *Il cammino dell'uomo: Perché siamo diversi dagli altri animali*, Garzanti, Milano, 2004 (*Becoming Human*, 1998).

Teilhard de Chardin, P. *Il fenomeno umano*, Queriniana, Brescia, 2008 (*Le phénomène humaine*, Editions du Seuil, Paris, 1955).

Tomatis, A. *L'orecchio e la vita*, Baldini & Castoldi, Milano, 1993.

Tonelli, G. *Cercare mondi*, Rizzoli, Milano, 2017.

Varela, F. J. *Un Know-how per l'etica*, Laterza, Roma-Bari, 1996.

Ventura, M. *Per una super-religione*, La Letture-Corriere della Sera, 10 luglio, 2016.

Visetti, Y.-M. Construtivismes, émergences: une analyse sémantique et thématique, *Intellectica*, **39**, 229–259, 2004.

von Balthasar, H. U. *Lo sviluppo dell'idea musicale*, Glossa, Milano, 1995.

Vrobel, S. *Fractal Time: Why a Watched Kettle Never Boils*, World Scientific, Singapore, 2011.

Yockey, H. P. *Information Theory and Molecular Biology*, Cambridge University Press, Cambridge, 1992.

Wald, G. The origin of life, *Sci. Am.*, **191**(2), 44–53, 1954.

Zhabotinsky, A. M. Periodic processes of malonic acid oxidation in a liquid phase, *Biofizika*, IX, Moskva, Nauka, 1964.

Author Index

Subject Index